ROUTLEDGE LIBRARY EDITIONS:
URBAN PLANNING

Volume 21

URBAN PLANNING
UNDER THATCHERISM

URBAN PLANNING UNDER THATCHERISM

The Challenge of the Market

ANDY THORNLEY

Routledge
Taylor & Francis Group

LONDON AND NEW YORK

First published in 1991 by Routledge

This edition first published in 2018
by Routledge
2 Park Square, Milton Park, Abingdon, Oxon OX14 4RN

and by Routledge
711 Third Avenue, New York, NY 10017

Routledge is an imprint of the Taylor & Francis Group, an informa business

© 1991 Andy Thornley

British Library Cataloguing in Publication Data
A catalogue record for this book is available from the British Library

ISBN: 978-1-138-49611-8 (Set)
ISBN: 978-1-351-02214-9 (Set) (ebk)
ISBN: 978-1-138-48822-9 (Volume 21) (hbk)
ISBN: 978-1-138-48999-8 (Volume 21) (pbk)
ISBN: 978-1-351-03626-9 (Volume 21) (ebk)

Publisher's Note
The publisher has gone to great lengths to ensure the quality of this reprint but points out that some imperfections in the original copies may be apparent.

Disclaimer
The publisher has made every effort to trace copyright holders and would welcome correspondence from those they have been unable to trace.

Urban Planning under Thatcherism

The challenge of the market

Andy Thornley

London and New York

First published 1991
by Routledge
11 New Fetter Lane, London EC4P 4EE

Simultaneously published in the USA and Canada
by Routledge
a division of Routledge, Chapman and Hall, Inc.
29 West 35th Street, New York, NY 10001

Laser printed from author's disks by
NWL Editorial Services, Langport, Somerset, England
Printed and bound in Great Britain by
Mackays of Chatham PLC, Chatham, Kent

British Library Cataloguing in Publication Data
Thornley, Andy
   Urban planning under Thatcherism: the challenge of the market
   1. Great Britain. Town planning. Political aspects
   I. Title
   711.40941

ISBN 0–415–05538–5
ISBN 0–415–05669–1 pbk

Library of Congress Cataloging in Publication Data
   Urban planning under Thatcherism: the challenge of the market/
Andy Thornley.
      p.   cm.
   Includes bibliographical references.
      1. Urban policy—Great Britain.   2. City planning—Great Britain.
   3. Great Britain—Economic policy—1945–   4. Conservatism—Great
   Britain.   5. Great Britain—Politics and government—1979–
   I. Title.
   HT133.T56   1990   90-8424
   307.1′216′0941—dc20   CIP

Contents

Contents

Acknowledgements

The ideas in this book have been maturing for as long as Thatcherism itself. I have therefore benefited from discussions with more people than can be acknowledged here. I would however like to express my gratitude to my colleagues and students at the Polytechnic of Central London for providing me with continued stimulus throughout the project. I would also like to thank the following people for their invaluable comments on all or parts of the book: Patsy Healey, Patrick McAuslan, Beverley Taylor, Cath Taylor, Robin Thompson and Bob West. In particular I would like to express my special gratitude to Derek Diamond for his continual support, encouragement and comment.

Chapter one

Introduction

Thatcherism has had a major impact on British society. There is no doubt that in some fields, such as education, housing or local government finance, Mrs Thatcher has pursued a long-term strategy to overhaul previous traditions and practices. This book explores the influence of Thatcherism on planning. Ministers are often stating their support for the planning system while simultaneously introducing initiatives such as Urban Development Corporations or Simplified Planning Zones, which undermine it. This book shows that these initiatives are not minor experiments but amount to a re-orientation of the planning system through a strategy of consistent incremental erosion. It explores the way planning has had to respond to the challenge of a more market-oriented ideology. As alternatives to Thatcherism become an increasing matter of debate it is important to gain a thorough understanding of what has happened since 1979. The future has to build on the foundations of the past.

A major element of the ideological rhetoric of the 1979 election campaign was an attack on post-war Welfare State values. The comprehensive town planning system was established in this post-war setting – does this mean that an attack on the values of this period undermines the basis of planning? The Nuffield Report on Town and Country Planning (1986) concluded that a variety of views on the proper role of planning existed and that, as a result, planning cannot place itself outside political controversy. The report identified trends in recent years that have led to 'the planning system being harnessed to ideological ends' (1986, p. 30). This has created more uncertainty than was experienced under the 'post-war' consensus. This book examines the implications for planning of the collapse of this consensus and the rise of Thatcherism.

In her much-quoted comment, Mrs Thatcher has claimed that 'there is no such thing as society; there are only individuals, and families'. Under Thatcherism the language of politics has shifted from 'public good' to 'individual choice' and 'entrepreneurial flair'. Bureaucrats and professionals, whether they be teachers, social workers or even solicitors, are not held in particularly high regard. The implications of these individualistic and anti-bureaucratic attitudes for the role of professional planners are considerable. Under the post-war consensus planners usually saw their role as guardians of the public or community interest but this goes against the grain of Thatcherism. A study of the interrelationship between the political ideology and practice of the Thatcher governments is needed to illuminate such issues. There is much theoretical debate about the exact nature of Thatcherism and this also needs to be taken on board. Before outlining the approach taken in this book some comments will be made on the reactions of planners to the advent of Thatcherism.

In many policy fields there have been strong reactions condemning the adverse effects of Thatcherism. However, planning has always been closely linked to market processes through its regulatory function. The attack on state intervention implicit in Thatcherism has revived the longstanding debates over the appropriate role of planning versus the market. Confusions over the purpose of planning reflect its position caught in the ambiguity between the egalitarian aims of the Welfare State and the individualistic attitudes implicit in the capitalist market economy. Given this position it is not surprising that the reactions of planners to Thatcherism have been varied and often confused. Some have tried to retreat into an apolitical and technical mode but this has proved difficult to do in an increasingly politicised environment. Others have swum with the tide and adopted the dominant values of central government, often joining or setting up private consultancies. Then there are those who have retained a belief in some kind of community or social purpose and have sought employment in the ever-diminishing sympathetic organizations.

The impact of Thatcherism has been interpreted in a number of ways. Although it creates rather artificial distinctions, these interpretations can be broadly divided into three categories emphasising either continuity, consolidation or change. The first category points out that despite the rhetoric very little has changed in practice. Here the emphasis is on the value of planning to society and the constraints that prevent any major overhaul of the planning system. The second category suggests that although the post-1979 situation may create changes in planning these simply reinforce and consolidate aspects that have always existed and therefore the changes are only a matter of increased emphasis on certain attributes of planning. The third category claims to identify certain features since 1979 that are significantly different from those of the past, thus generating a new position for planning.

The first kind of interpretation was widely held in the early years of the Thatcher administration when it was thought that much of the ideological rhetoric would be abandoned when the new government faced up to the difficulties and limitations of power. Such a view was reinforced by the government's seemingly incremental and pragmatic approach to changes in planning. There was no major new planning Act to overhaul the system and the Local Government, Planning and Land Act of 1980 was made up of odds and ends. The ministerial statements stressed that the government was supportive of the planning system and this therefore lent weight to the 'continuity' interpretation. Changes seemed to be confined to making the system more streamlined and efficient rather than changing its role (see Healey, 1983, p. 269). Most early writing assumed that the more radical measures would be watered down under the pressure of government (Cox, 1980).

The 'consolidation' view accepts that there has been a change since 1979 but emphasises that this change is an extension and reinforcement of past trends. The central theme is that the economy has always been a capitalist one and hence market processes have always dominated. Planning has inevitably assisted and reinforced these market processes. Thus the nature of the planning system since 1979 is basically the same as before although there may have been some increases in the power of the development industry. In other words the role of planning in aiding the market has been consolidated. Whereas before 1979 this role may not have been popular and was often obscured, since that date its true nature has been exposed and made 'acceptable' (Griffiths, 1986).

Another dimension of the 'consolidation' view is that the beneficiaries of the planning system have always been the middle classes and therefore a Conservative government will see no value in dismantling it. Lawless (1983) suggests that the changes brought about in Thatcher's first term had little effect. The reason he gives for this is that planning practice in the past has favoured suburbia, owner occupation and the middle classes and that Thatcher will want to continue this support. This line of argument is also pursued by Reade (1987) who sees planning as having kept up market values in expensive residential areas, kept working-class housing and other undesirable development out of 'commuter country' and protected rural villages. He concludes that the attack on planning is purely symbolic and an attack on the 'word' rather than the practice. Writers adopting the 'consolidation' position often refer to a shift in the interests served by planning from support for conservation before 1979

3

to greater support for development interests. Thus these writers generally accept that within the 'consolidation' of planning purpose there have been shifts in the relative strengths of different interests (Healey, 1986; Reade, 1987).

The third perspective gives more emphasis to the new dimensions brought about by Thatcherism. It suggests that the movement towards a greater reliance on the market carries with it other significant changes. McAuslan (1981, 1982) highlights some of these changes in his analysis of the legislation and circulars of the early years of the Thatcher administration. He suggests that since 1979 we have witnessed a shift in the whole approach to planning. The post-war period was dominated by what he calls the 'bargaining policy-oriented model of planning' epitomised by a process of mutual negotiation between planners and developers. This model incorporated an important role for a policy framework. Since 1979 this has been superimposed by a 'limited physical control model' whose primary aim is to facilitate development and which excludes social objectives. He sees these two models continuing alongside each other and creating tension because local authorities favour the bargaining model while central government favours the physical control model. He also sees the actions of central government leading to increased centralisation and reduced opportunities for participation. Thus from this perspective the changes since 1979 are not just a consolidation of the market supportive role of planning but also a significant shift in the ability of local authorities and the general public to take part in the development process. It could be said that this represents the separation of planning from its roots in the post-war Welfare State consensus which embodied an acceptance of social criteria in decision-making.

The degree of coherence in the changes to planning since 1979 also requires discussion. Healey refers to 'the incoherent and indiscriminate grapeshot which the 1979 government directed at the planning system' (1983, pp. 263–4). Any consistency of approach is limited by the contradictory pressures on the government. Healey gives an example of these contradictions when she points to the desire to relax regulations but at the same time the need to intervene to support certain development in a period of economic crisis. Perhaps the most coherent programme for change comes from groups, such as the Adam Smith Institute, that have been lobbying the government. The extent to which such ideas have influenced government and provided a longer-term hidden agenda behind the incremental initiatives will be another question addressed in this book. Evidence that some senior civil servants are working to an overall direction sometimes surfaces, for example in the suggestion that the simplified planning regimes could replace the 1947 development control system.

The extent to which the Thatcher governments have generated a radical change in the approach to planning and the degree of coherence embodied in the changes are central issues to be explored.

The approach adopted in the book

So the rise of Thatcherism has had the effect of highlighting the relationship between political ideology and planning. But, as Healey has said, the 'debates about the present state and future possibilities for land policy are still characterised by simplistic connections between political philosophies and legislative tools' (1983, p. 264). The problem with the current literature on changes to planning over the last decade is that it does not provide a thorough analysis of the new ideological context. An improvement in the quality of understanding about the relationship between Thatcherism and planning can be obtained only through a detailed investigation of both the ideology and its implementation. The approach adopted in this book will be outlined.

First there is the question of focusing on ideology. Changes to planning are caused by a complex interaction of economic, political, ideological and social factors and there is a vast body of theoretical literature exploring the relative importance of these factors. The work which emphasises ideology often draws on Habermas or Gramsci (for a review of these debates see for example, Abercrombie, *et al.* 1980; S. Hall, 1988; for an application to planning, Hajer, 1989). There is not the scope here to address these interesting theoretical debates but brief mention is made of how ideology is viewed in relation to the broader picture.

The most compulsive force operating in society in recent times is that of economic crisis. This had led to the need for a reorientation of the economy and the related structure of power in society. However such a reorientation requires both the backing of the population, or at least a sizeable proportion of it, and the tools to implement the necessary changes. A political ideology which is conducive to the changes required in the economy needs to be developed and 'sold' to the population. This new ideological context allows the appropriate reorientation of the role of the state and the passing of new legislation. Once these facilitating moves have been made on the ideological and legislative terrain then the opportunities are open for economic interests to restructure and changes to be made to power relations. For example the inter-war economic crisis was 'resolved' after the Second World War by a new approach to economic affairs involving an interventionist stance by the state but this had to be preceded by considerable debate and propaganda over the acceptability of economic planning. In the context of 1979 one of the obstacles to a radical reorientation of economic strategy was the value system that had evolved under the post-war Welfare State and there was a need for an ideological attack on this value system as a precondition of the economic restructuring. Thus economic forces play the dominant role in determining the nature of the crisis and shaping the parameters of the political response. However, ideology provides an essential mechanism for the implementation of this response and a vehicle for shifting the ground rules within which economic interests operate. It should also be mentioned that the elements of any ideology have usually been in existence for a long time but come forward only when they can play a significant role in relation to other material conditions. Ideology is therefore seen as having an important role in influencing change. It is particularly important in establishing a new framework, such as that provided by planning legislation, within which subsequent action can take place.

Planning is a very broad and vague field and so it is necessary to draw boundaries around the topic if a detailed analysis is to be made. The focus in this book is the planning system, that is the legislation, Circulars and other material that determine the basis of the development plan and development control procedures. The book examines the content of these documents in order to assess the way in which they shape the scope, purpose and form of planning. It must be stressed that this emphasis on the legislation and procedures of the planning system can be only a partial investigation of planning activity as a whole. It is not the intention in this study to explore how far the changes in legislation and procedures have affected the operation of planning on the ground. To pursue this dimension research would have to be undertaken, for example, on the way in which the legislative and procedural changes have altered the manner in which planners operate from day to day, especially in their relationship with the private sector, and case studies mounted on the implementation of development schemes (see for example, Healey *et al.*, 1989; Brindley *et al.*, 1989).

The focus on the planning system means that many aspects related to planning practice have been excluded. These include legislation on other subject areas such as housing or transport and on specific smaller areas within the system such as inner city grants or Green Belts. Other dimensions not incorporated are regional planning, purely rural issues such as controls on agriculture or wildlife, and initiatives that apply only to Scotland. Having made this initial distinction it has to be accepted that it is difficult to draw exact boundaries. For example reference is made in the book to the debate over the allocation of land for housing and considerable coverage is included on Urban Development Corporations. The criteria for drawing such material into the study is its impact on the mainstream planning system itself rather than a wish to enter the particular debates on housing land or inner city policy. Pressures for more housing land and relaxation of controls in the Green Belt are issues that have created considerable controversy in the last decade, and have been very well discussed elsewhere (see Elson, 1986; Rydin, 1986). The implications of these topics for Thatcherism will be encompassed in the conclusions of this book.

Thus, within the limits set, the approach is to explore in detail all the changes that have been made to the legislation of the planning system. These changes have not been dramatic ones and there is therefore a need to draw together the implications from a wide range of smaller initiatives. Patterns of repeated phenomena are detected and related to the main themes of Thatcherism. Clear trends are identified from this process and conclusions made about the impact of the ideology. These conclusions can be applied to a wider range of planning issues and future developments and at the end of the book some broader observations are made.

The structure of the book

The aim of the first part of the book is to identify the precise nature of Thatcherism and draw out those aspects that are likely to have the greatest impact on planning. In the book the label 'Thatcherism' has been used to describe the ideology put forward by the Thatcher governments or particular ministers and discussed by political analysts. The label 'New Right' has been used to describe the broader range of ideas from which the government draws its inspiration, for example those emanating from academics such as Hayek or bodies such as the Adam Smith Institute.

The study of Thatcherism has to be put in historical context to judge the degree of change and to assess the extent to which the period can be regarded as a radical departure from the past. Chapter Two provides this context giving particular attention to the establishment of the post-war consensus. This consensus has been the subject of attack from within the Thatcherist ideology and so it is important to establish its nature and relationship to planning. As already mentioned, the Nuffield Report (1986) shows how different political ideologies imply different attitudes to the 'proper' purpose of planning. The chapter therefore also reviews the debates over the purpose of planning in the post-war period as a bench-mark against which later to judge the significance of Thatcherism.

Chapter Three sets out the principal characteristics of Thatcherism, discusses the extent to which it can be regarded as coherent and the extent to which it represents a break with the conservative thinking of the past. In exploring these issues the debates and controversies over the interpretation of Thatcherism will be reviewed. Chapter Four continues this process through an examination of how the state is viewed by influential New Right academics and the way their ideas have been taken up in the political arena. As the planning system has always operated as a state activity that intervenes in the processes of land use and development, attitudes regarding the acceptability of state intervention are crucial to understanding potential changes. Particular attention is given to the views of Keith Joseph and Margaret Thatcher herself.

The analysis of ideology and the investigation of the planning legislation have to be synthesised. The linkage is developed in a number of ways. The coverage of planning purpose, already mentioned, is one way ideology and practice are brought together. In addition a framework of the key attributes of the Thatcherite ideology is constructed drawing upon the analysis in Chapters Three and Four. This framework is then used to structure the subsequent discussions about the changes to planning. The themes drawn from the analysis of the political ideology are thus integrated into the detailed accounts of planning legislation and procedures and the concluding chapter focuses on the integration of the various strands.

Chapter Five provides another bridge between discussions of political ideology and the details of the planning system. It reviews the work of authors who have written about planning from a New Right perspective. Their views challenge those embodied in the post-war consensus analysed in Chapter Two. The material covered in Chapter Five has been chosen because of its explicit influence on the Thatcher governments. It includes important influences from the USA, the work of organizations such as the Institute of Economic Affairs, the Department of Land Economy at Cambridge, the Adam Smith Institute and certain key individuals.

The second half of the book explores in detail the changes to the planning system during the first decade of Thatcherism focusing on the legislation, Circulars and ministerial statements. However, it is not considered sufficient to look at modifications to the existing planning system because significant implications also result from initiatives which adopt alternative mechanisms or procedures. The changes since 1979 are therefore analysed under the three headings of modifications to the existing system; avoidance or by-passing of the system; and initiatives that have the potential to replace the system.

Chapters Six and Seven examine the modifications in detail. First, the development plan part of the system is investigated, from the early ministerial comments on particular structure plans through the implications of the Local Government, Planning and Land Act, 1980, the White Paper *Lifting the Burden* (Department of Environment (DOE), 1985a) and Unitary Development Plans, to the White Paper on *The Future of Development Plans* (DOE, 1989). Chapter Seven covers the development control side and particular attention is given to Circular 22/80 (DOE, 1980b), planning gain, planning conditions, control of industry and small business and changes to the General Development Order and Use Classes Order. An assessment is made of the cumulative effect of all these modifications.

Chapter Eight is based upon the concept that certain initiatives provide the opportunity to by-pass the formal planning system. As these initiatives become more frequently used so the impact on the planning system becomes more significant. Three initiatives are identified: architectural competitions, Special Development Orders and Urban Development Corporations. The last of these is given particular attention as the application of this initiative has been greatly extended. The planning approach in Urban Development Areas is analysed in detail. Chapter Nine focuses on simplified planning regimes as instigated in Enterprise Zones and Simplified Planning Zones. Again the procedures adopted in these simplified regimes are compared to those in the 'normal' planning system. It is suggested that these simplified regimes provide a potential alternative to the existing system of planning regulation based on the 1947 Town and Country Planning Act.

The final chapter draws together the various threads of the study synthesising the analysis of the political ideology and the detailed examination of legislative and procedural changes. Considerable coherence can be detected in the pattern of changes and the net result is a significant reorientation of the principles and procedures of the planning system. However, contradictions and weaknesses in the application of the ideology are also identified. Finally some thoughts are presented on the implications of a decade of Thatcherism for the future of planning.

Chapter two

The post-war consensus and the purpose of planning

Thatcherism needs to be considered in relation to the past. The rhetoric of Thatcherism constantly refers to the mistaken path taken by post-war governments. This chapter looks at this post-war period concentrating upon those aspects that help illuminate the nature of Thatcherism. The aim is to provide enough background to understand the degree of change that Thatcherism has brought to the planning system. A significant feature of Thatcherism is its attack on the post-war consensus which is a precursor to the reconstitution of the relationship between the market and state intervention. This feature provides a focus to the examination of the historical material. The first question that is addressed concerns the nature of this post-war consensus and the role adopted by the state in the period up to 1979. The relationship between planning and this consensus is then investigated.

The debates over the role played by planning in the post-war period are elaborated in order to provide a bench-mark against which to judge the significance of Thatcherism. Such an examination of the purpose of planning is not straightforward as neither the legislation nor statements from planning practitioners have made this purpose clear. The question is approached from several directions. The Thatcherist challenge has stimulated a consideration of the variety of possible objectives that can be pursued by planning. These are examined. The emphasis given to these different objectives throughout the post-war period is then explored focusing on the way in which any particular configuration of objectives relates to the role of the state and the market. Attention is given to the contradictions and conflicts in the role of planning.

The post-war consensus and the changing role of the state

This section examines the establishment of the post-war consensus and outlines the way in which the role of the state has been adapted throughout the post-war period. The question of the role of the state has generated considerable debate and it is not intended to review fully or analyse the literature on this subject (for a detailed discussion of theories of the state see for example, Dunleavy and O'Leary, 1987; Jessop, 1982; Offe, 1984; Urry, 1981; and for examples of the implications for spatial policy, Cooke, 1983; Clarke and Dear, 1984; Saunders, 1979, 1981). The aim here is to provide sufficient material to place the discussion of post-war planning in the context of the role of the state.

Three broad forces are identified as the reasons for the development of consensus. They are the inter-war experience, the effect of the war itself and the reactions of the political parties. The first of these factors provides the broad preconditions for the attitudes adopted during the war and the political response. The inter-war period was one of extreme crisis for the capitalist system with major economic problems and social unrest. This crisis contributed to the widely held view that after the war a new approach had to be found because there could be no return to the pre-war conditions. The foundations for this new approach had already been laid in the 1930s. Harris (1972) has charted the battle between the two strands in the ruling inter-war Conservative Party, the supporters of the free market and the supporters of state involvement, and he calculates that the latter group gained ascendancy within the party in 1931. He states the reasons for this change as the failure of the free-market-oriented policy of returning Britain to the Gold Standard, the need for a protectionist economic policy requiring a role for the state, and changes to the structure of firms towards larger companies resulting in less flexibility and a need for greater state-inspired stability. This mutual relationship between large firms and the state was to be strengthened in the war itself as the state used large firms to provide the needed reliability of goods for the war effort.

This shift in thinking towards a role for the state in economic planning is illustrated by the rise in the 1930s of groups and individuals pressing for this approach. An organization entitled 'Political and Economic Planning', made up of industrialists, businessmen and civil servants, was established in 1931 to lobby for a state role in economic planning. In 1936 Keynes published his *General Theory*. Two years later Macmillan published his book *The Middle Way* in which he called on all sections of society to back the new relationship between state and economic interests. He presents the 'new doctrine' in the following terms;

> This book is offered as a contribution towards the clearer formu-
> lation of the new ideas of society that have been slowly emerging
> since the political crisis of 1931. I hope it will be given sympathetic
> consideration by men and women of all parties who recognize that
> some new theory of social evolution must be conceived if we are
> to retain our heritage of political, intellectual, and cultural free-
> dom while, at the same time, opening up the way to higher
> standards of social welfare and economic security.
>
> (Macmillan, 1938, pp. 5–6)

Politicians of all parties were attracted to the idea of economic planning but as Deakin (1987) points out even at this stage the advocates
often supported the concept with differing objectives, some more
interested in the modernisation of industry, some the provision of
more efficient services and the health of the nation and some seeking
a fairer distribution of wealth. This coalition of interests meant that
in the 'great debate' on the merits of economic planning that took
place at the time under the threat of totalitarianism, the pro-plan-
ners were gaining support at the expense of their opponents, such as
Hayek (see his *The Road to Serfdom*, 1944). The war itself provided a
further jolt to thinking and action.

Titmuss (1950) in his official account of the social policy aspect of
the war talks about a social contract between government and the
people. He suggests that the deprivations of war, the impact of evacu-
ation, the greater awareness of poverty and social differences all led
to a mood of greater collectivism and desire for equality. In response
government had to present a vision of a better society that overcame
the inter-war problems. The Beveridge Report and the popular interest it generated can be viewed in this light. It could be said that one
of the results of the war-time experience was to consolidate the demand for the social rights of citizenship to supplement political
rights (Marshall, 1964). However, as indicated above, many of the
ideas for this 'new society' had already been devised before the war
and thus the impact of the war itself as a major reason for the establishment of the Welfare State needs to be kept in perspective (for a
fuller treatment of these issues see for example, M. Bruce, 1968, or
Fraser, 1973).

The third dimension in the analysis is the role of the political parties. The performance of the Coalition government is crucial here.
As already mentioned, there was cross-party support before the
Second World War for economic planning and the resultant role
for the state. This carried through to the Coalition government of
1940–45. Kavanagh (1987) suggests that the Cabinet Committee on

Reconstruction set up in 1943 was particularly important in reflecting the convergence between the political parties. This committee reached agreement on a wide variety of policies covering the National Health Service, regional policy, full employment, social insurance and housing. The pragmatic acceptance of the mixed economy and the Welfare State continued after the war. Although it fell to the Labour Party to implement the required measures these did not meet with opposition in principle. A political consensus was evident for a larger public sector and a reduced role for the market. On the Conservative side this consensus was facilitated by the moves within the party, led by Butler, towards a 'New Conservatism' which consolidated the support for an interventionist state (see for example, Deakin 1987).

The resulting role of the state covered a number of interconnected areas, for example, state control of basic services and industries, provision of social welfare and the requirement for high public expenditure and taxation. The interconnected nature of this package was clearly seen at the time and viewed as an important contrast to the inter-war situation. The social expenditure was seen as necessary for engendering the stability needed for the growth of the economy. Simultaneously this growth was necessary to support the additional expenditure. In a speech in the House of Commons in 1946, on the National Insurance Bill, Attlee described this relationship between the government's social measures and the economy of the country. He explained that the Bill was part of the Coalition government's full employment policy:

> We now recognize that, to allow, through mass unemployment or through sickness, great numbers of people to be ineffective as consumers is an economic loss to the country. We all now hold the view that we must maintain purchasing power and must have a proper distribution of purchasing power among the masses of the people. The old idea that this could only be done through wages and profit, and not by collective provisions of this sort, is now, I think, dying . . . it is interesting to see how far, in quite a short time, we have travelled from the conception of the panic cutting-down of the purchasing power of the masses, which was employed as a means of dealing with the abundance crisis of 1931. We now realise we have the backing of experienced administrators and of the economists that it is necessary, in this matter, to have a degree of planning.
>
> (quoted in Gregg, 1967, p. 45)

This speech has been quoted at length because in addition to illustrating the interconnected nature of the post-war social and economic policies it shows again the influence of the inter-war crisis and the close co-operation of experts, administrators and politicians at the time.

Thus it can be seen that the consensus developed as a means of resolving basic economic and social problems of the inter-war period which were accentuated by the war itself. The 'resolution' consisted of generating a package that integrated both economic and social dimensions and utilised the state as an important instrument in creating and maintaining this package. The various interests, while all agreeing in the broadest terms that such a package was necessary, had different hopes and aspirations regarding the benefits of the arrangement.

The fragile nature of this alliance of interests was to become evident later and in particular made such a consensus vulnerable to the attack by Thatcherism. One aspect of this vulnerability was the elite nature of the enterprise. Notwithstanding the popular sentiments expressed during the war the post-war approach to government was an elite one. The Welfare State was developed along Fabian lines, through a partnership between government and expert administrators. An attempt was made to use the state to redistribute wealth throughout society. It was assumed that this redistribution could be carried out quite painlessly and that all the necessary expert knowledge was available which just needed to be marshalled to the cause. Two consequences of this approach are worth mentioning. First, as Kavanagh (1987) points out, the consensus was mostly at the level of elites, for example Members of Parliament and experts, rather than at the popular level. It could be said that the general public were quite happy to go along with this approach so long as it created material prosperity. As will be shown later, once this prosperity collapsed, the criticisms and questioning began and opened up the scope for the Thatcherist challenge.

The second consequence of the elite approach was the importance and power it gave to experts and bureaucrats. This had the effect of depoliticising decision-making and led to the characterisation of the period as one of the 'end of ideology'. The dominant group were the Keynsian economic experts, those who could orchestrate the consumer boom for Macmillan or the technological revolution for Wilson. The emphasis was on the efficient administration of the mixed economy and the establishment of the economic growth on which the whole strategy depended. The classic statement of Labour's allegiance to this approach was Crosland's influential book *The Future of Socialism* (1956).

A further related feature of the consensus approach to government was its corporatist nature. There is some debate over the concept of corporatism and the term is used here to mean the approach whereby government involves itself in consultation and bargaining with representatives of the major economic interests such as industrialists and trade unions (for a fuller account of corporatism see for example, M.L. Harrison, 1984; Cawson, 1986; Reade, 1987). As seen above this form of co-operation was starting in the inter-war period with the changes in economic structure and the development of support for a state role in economic planning. Such an approach was to be a feature of the post-war consensus and was to be developed further in the 1960s with special bodies and organizations set up for the purpose of co-operation. Such an approach was still central to Heath's government after its U-turn, while Wilson's 1974 government laid great stress on its ability to manage economic interests and devise a 'Social Contract'. One of the implications of the corporatist approach was to reinforce the power of experts and bureaucrats.

It is generally agreed that the essential features of the post-war consensus were maintained until 1979. This included inter-party acceptance of the mixed economy, the welfare role of the state and a corporatist approach to government. The ability of such an approach to generate economic growth ensured that the potentially conflicting elements in the consensus were contained (see Held, 1984). However, this economic precondition of the consensus proved more and more difficult to achieve, eventually exposing some of the inherent weaknesses in the alliance of interests and its elite nature. The final part of this section will examine the cracks that developed in the consensus in its later years.

Opposition to state intervention from within the Conservative Party, having been defeated in the early 1930s, did not completely disappear. A reassertion of this opposition came from Powell in the 1960s when he made a consistent and repeated attack on the activities of both political parties and, based upon fundamental principles, propagated the merits of a free-market approach (for details see for example, Gamble, 1974; Kavanagh, 1987). These ideas did not fall on fertile ground at the time but were taken up in the late 1960s to form the basis of Heath's 'Selsdon' approach. However, the subsequent U-turn led to a re-establishment of the consensus principles which was only severely questioned in the Conservative Party when they lost the 1974 election. At this point there was a major revival of free-market thinking spearheaded by Joseph and later Thatcher.

As already stressed, the maintenance of the consensus depended upon economic growth which enabled all groups to be satisfied and generated support for the strategy. However, the beneficial conditions for growth, which existed temporarily in the 1950s, ceased and the inherent weaknesses of the British economy reasserted themselves. As a result the growth of the economy became more and more difficult to sustain. There is considerable literature and controversy over the nature of this decline and the underlying reasons for it (see Coates and Hillard, 1986, for a coverage of these debates). Increases in unemployment and inflation were experienced and the state had difficulty in financing its activities. The resultant cut-backs in public expenditure created resentment and threatened social support for the consensus. As the political parties in government struggled to continue with a consensus approach in the harsh economic climate, yet failed to produce results that met public expectations, so the middle ground of politics lost public credibility (Held, 1984; Leys, 1983).

The events of 1968 represent most dramatically the breakdown of the consensus in its early stage. Here was a challenge to the acceptance of government authority and a questioning of the values and approach of governments. This was expressed, for example, in terms of the Vietnam War and the importance of peace, and university decision-making structures and the importance of participation. Such challenges have been described as 'a revolt against the pragmatic, technocratic values of advanced industrial society' (Eccleshall, 1984, p. 12). From this movement arose many demands that continued and developed in the subsequent period, such as those advocating democratic principles, women's, lesbian and gay rights and a resistance to racism. As will be seen later a backlash also developed with an emphasis on the sanctity of the family and the need for discipline in society. As well as indicating a challenge at the level of values, these events and the tensions and antagonisms that developed indicate the failure of the consensus to contain all the diverse elements within its ambit. With a shrinking national economic cake the conflicts between groups developed as they fought to maintain their share.

It was during this period too that the post-war Welfare State approach to poverty was questioned. In what was labelled 'the rediscovery of poverty' it was realised that within all the affluence and prosperity of the post-war years there were many people who were still suffering. The redistribution approach had not reached everyone. As a result numerous special commissions were set up to examine different aspects of the problem, for example housing, education, or social services. The recommendation common to all their reports was the need for a selective, area-based, positive discrimination approach with greater participation, to supplement the universalist Welfare State.

One dimension of this questioning of values is of particular interest. The dissatisfaction with decision-making processes and the reinforcement of democratic rights led to an attack on corporatism with its inherent secrecy involving the by-passing of Parliament and local councils. This attack also implied a questioning of the power of experts, leading politicians and bureaucrats. The opposition came from all shades of political opinion. There was the move for greater citizen involvement and the acknowledgement of the importance of community and voluntary groups. This was developed later into attempts to create greater democratisation of local government through mechanisms such as popular planning and decentralisation. Within the Labour Party it was appreciated that the remote bureaucratic nature of the Welfare State created problems (for examples of these arguments see London Edinburgh Weekend Return Group, 1979; Gyford, 1985). However, as will be discussed in detail later, there was also a strong challenge from the right to the corporate, bureaucratic style of government. In this case the answer was to be found in an emphasis on the liberation of consumer choice and personal responsibility.

The inability to maintain economic growth thus exposed social conflicts and led to a disillusionment with government. It also led to a theoretical questioning of the role of the post-war state. Attlee in his speech to Parliament, quoted above, indicated how the post-war consensus approach of the state combined both social and economic functions. Much of the theoretical debate on the breakdown of the consensus concentrates on this dual function of the state, promoting social harmony and welfare while also aiding the economic growth process. A major feature of the developing crisis is seen to be the need for the state to pursue its welfare function to provide stability while being unable to create sufficient surplus in the economy to continue paying for this increasingly expensive role (O'Connor, 1973). Held (1984) identifies two strands to the theoretical debates on this crisis of the state: 'government overload' theories such as those of Brittan and King which draw upon a pluralist approach, and 'legitimation crisis' theories such as those of Habermas and Offe drawing on Marxism. In analysing these different strands Held shows that, although there are differences between them due to their pluralist or Marxist origins, they share a considerable area of common ground. This common ground will be outlined as it incorporates the issues facing the state in recent times and therefore provides a useful background to the consideration of Thatcherism.

Both theoretical strands see that a major aspect of the crisis is the inability of the state to continue generating public acceptance of its activities. Both see growing demands and claims being placed on a state that has less and less power to meet such demands. However, the theories proffer different explanations for these trends. Overload theorists refer to increased affluence creating higher aspirations accompanied by a loss of deference for government. Competing groups put more and more pressure on governments which try to appease all these demands to ensure political survival. However, spiralling costs mean that all demands cannot be satisfied, fuelling further disillusionment. Legitimation crisis theorists refer to the need for the state to undertake greater and greater activities as it tries to carry out its necessary functions of economic stimulation and social harmony. The cost of the increasing administrative structures is a drain on the economy which becomes more difficult to control with rational policies. The imperative of ensuring economic growth and profit margins creates a cut-back in welfare-oriented activity thereby threatening the state's legitimacy.

As Held (1984) points out, the analysis does raise some questions. First, to what extent does this deference to, or legitimacy of, the state really exist? It has been noted above how the post-war consensus can be viewed as based upon popular acceptance because of the material rewards it generated in a period of affluence. Thus it could be said that the allegiance to the state is not one of principle but conditional on the state creating the material results. What has happened then is not necessarily a change in attitude to the state but a change in the state's ability to come up with the goods. A second comment on the theories is their assumption that the power of the state is being eroded. However, this can be viewed not as a reduction in state power but a shift in its approach. For example Held refers to the strategy of displacement, transferring the blame for poor performance from government on to particular groups in society or some external factor beyond the control of a national government. In this way government does not take on the responsibility for producing the acceptable economic and social results but relies instead on traditional notions of authority. Other writers have noted the shift in the state's role to an increased emphasis on law and order. These are issues that will be pursued later in detail when analysing Thatcherism. Meanwhile it is necessary to examine town planning in this period in the context of these discussions of the role of the state.

The purpose of planning

The variety of aims, or purposes, attributed to planning and the way these different purposes have been employed historically will be discussed. It is not the intention here to discuss the particular mechanisms or means by which planning attempts to achieve its objectives, such as containment, reduction of congestion or revival of inner cities (see Hall *et al.*, 1973). Instead the aim is to concentrate on the broader question of the role of planning in society and in particular its relationship to the market/intervention debate.

The outline of the possible purposes that can be attributed to planning draws upon two recent accounts. First, the Nuffield Foundation Report (1986) which was itself based upon submissions from a wide range of planning organisations and individuals. The Report concluded that many of the differences of view about the planning system can be related to underlying differences over the fundamental question of why there should be planning at all. The findings of this report are supplemented by the review of Klosterman (1985). His account is useful because it introduces an American dimension which, as will be seen later, is an important influence on the New Right's challenge to planning. The following range of planning purposes can be identified:

1 to improve the information available to the market for making its locational choices;
2 to minimise the adverse 'neighbourhood effects' created by a market in land and development;
3 to ensure the provision of any 'public goods', including infrastructure or actions that create a positive 'neighbourhood effect', which the market will not generate because such activity cannot be rewarded through the market;
4 to ensure that short-term advantage does not jeopardise long-term community interest;
5 to contribute to the co-ordination of resources and development in the interest of overall efficiency of land use;
6 to balance competing interests in the use of land to ensure an overall outcome that is in the public interest;
7 to create a good environment, for example in terms of landscape, layout or aesthetics of buildings, that would not result from market processes;
8 to foster the creation of 'good' communities in terms of social composition, scale or mix of development, and of a range of services and facilities available;
9 to ensure that the views of all groups are included in the decision-making processes regarding land and development;
10 to ensure that development and land use are determined by people's needs not means;
11 to influence locational decisions regarding land use and development in order to contribute to the redistribution of wealth in society.

This list of purposes has been used, selectively, in different ways by different people, sometimes to support planning, sometimes to attack it, and sometimes to try to explain it. Statements of purpose can fulfil the function of providing practitioners with a justification for their existence, an operating ideology. It can also be seen that the purposes are often contradictory and involve different attitudes to the market mechanism and the role of the state (see Healey, 1983). This therefore raises the question of how the contradictory elements can co-exist at any one time and how they relate to the broader political ideology. This book is concerned with the way in which the changes to planning legislation since the advent of a government with a Thatcherist ideology have affected the purpose of planning. First, a background is provided covering the way these issues of contradiction have been resolved in the post-war period.

There is another dimension to this question of planning purpose. This is the assessment by analysts and commentators of the purpose of planning as seen not in terms of what practitioners believe they are doing, but in terms of planning's perceived effects and contribution to the social and economic functioning of the country. Such purpose may include unintended consequences that do not figure strongly in planners' own view of their role. Although this dimension will be covered in the review of the post-war period, the issues of the relationship between an 'operating ideology' and the 'real' function of planning in society or the question of disjuncture between intended purpose and actual results will not be considered in detail (for such discussions see P. Hall *et al.*, 1973; Hague, 1984; Ravetz, 1980; and especially Reade, 1987). The main issue that arises in debates over this disjuncture is who benefits from planning intervention. Most of the stated purposes listed above assume the benefits should accrue to the 'public as a whole' or, in relation to the redistributive purposes, to the poorer and less vocal sections of society. The critics would claim that such intentions have not materialised and the benefits have more often gone to the powerful and privileged members of society and hence planning has reinforced social differences. This has led some writers (for example, Reade, 1987; Ambrose, 1986) to suggest that the overall purpose of planning is therefore to provide a mystification of reality. This mystification suggests that market forces and dominant interests have been checked and modified whereas in practice this has not been the case. Thus planning contributes to legitimising the inequalities of a market system by providing a pretence of government intervention in the 'public interest'. For example Reade concludes that 'the legitimacy of the property development industry and its associated financial institutions is maintained . . . by having a "planning system"' (1987, p. 67). The extension of this argument is that the function of planning is to contribute, through its legitimising function, to the conditions necessary for the maintenance of a capitalist market system (see for

20

example, Hague, 1984; Ball, 1983). Stemming from these discussions a further purpose of planning can be identified along the following lines:

12 to contribute towards the maintenance of the capitalist system and in particular provide an ideology of intervention in the 'public interest' while, in reality, supporting dominant economic interests.

The following sections interpret the way in which the historical circumstances of the post-war period have given particular substance to the question of planning purpose.

Planning as the child of consensus

This first section in the review of post-war planning concentrates on the period in the 1940s when the comprehensive planning system was established. The influence on the planning system in its formative years of the broader post-war consensus on the role of the state is explored. There are numerous good comprehensive accounts of the forces and ideas that contributed to the establishment of the system in the 1940s (for example Ashworth, 1954; Hebbert, 1977) and the intention here is to draw out some major issues that relate to the purpose of planning and the establishment of consensus. As Hebbert has shown there were many diverse interests that contributed to the establishment of the planning system and it is significant how this diversity, and its inherent conflicts, were moulded into an acceptable compromise.

The planning system was established with a broad cross-section of support from all parties and from different interest groups (see Cullingworth, 1975, for details). The contemporary observation made by Robson in his book, *The War and the Planning Outlook*, was typical of the period:

> In the two years that have elapsed since the outbreak of war an extraordinary change has taken place in the mental climate of this country on the question of town and country planning. For the first time the planning idea has suddenly become accepted as inevitable and necessary by large numbers of people belonging to all political parties and all classes of society.
>
> (quoted by Hebbert, 1977, p. 131)

Backwell and Dickens (1978) suggest that the war conditions favoured certain economic interests, particularly manufacturing industry and agriculture, because of their importance to the war effort. After the war these interests were in conflict over the use of land as industry required new sites and greenfield locations while landowners often wished to maintain their agricultural activity in a climate which placed importance on the ability of the country to produce its own food. The acceptance of planning can be viewed in the context of the need for a solution to these conflicts.

Such a role of balancing interests was supported by popular enthusiasm for planning as a means of creating a new post-war society. As discussed above the experiences of the inter-war period and the crisis of the capitalist system generated a climate for a radical rethink of how society was organized. Totalitarian regimes in the inter-war period in Germany and the Soviet Union were demonstrating an efficient approach at a time when Britain was suffering acute depression and social conflict. This was the background to the broad debates on planning mentioned above. From these debates emerged the agreement that an increased role for the state was necessary and that this could be linked to a democratic system of government without resulting in the extremes of totalitarianism that Hayek and others predicted. Planning was to play its part in the new scheme for society as the following extract from Thomas Sharp's best-selling Penguin paperback of the time shows:

> It is no overstatement to say that the simple choice between planning and non-planning, between order and disorder, is a test-choice for English democracy. In the long run even the worst democratic muddle is preferable to a dictator's dream bought at the price of liberty and decency. But the English muddle is nevertheless a matter for shame. We shall never get rid of its shamefulness unless we plan our activities. And plan we must – not for the sake of our physical environment only, but to save and fulfil democracy itself.
>
> (Sharp, 1940, p. 143)

In 1941 *Picture Post* published a special edition entitled 'Plan for Britain' which covered all aspects of life from work to leisure and education and included a section by Maxwell Fry called 'The New Britain Must be Planned'. The function of this publication is clearly stated in the following extract from the Foreword: 'Our plan for a new Britain is not something outside the war, or something *after* the war. It is an essential part of our war aims. It is, indeed, our most positive war aim. The new Britain is the country we are fighting for' (*Picture Post*, 1941, p. 4). This role of planning in contributing to the maintenance of morale during the war is taken up by Backwell and Dickens (1978) and Ambrose (1986). They show how the War Cabinet was very aware of the role that planning could play in providing a morale boosting vision of the future but also how there was a certain waryness of making specific promises. Certainly during the period, in films supported by the Ministry of Information, in the press and in popular paperbacks, considerable public awareness of planning was generated.

Thus the planning system was established on the basis of a consensus supported by what Donnison and Soto call a 'loosely knit alliance ... of diverse groups with common interests in the orderly management of land' (1980, p. 3). They describe these groups as

> the public health movement founded by Chadwick and the great medical officers of the previous century; the philanthropic and municipal housing movements, and the campaigners for new towns; farmers, country gentlemen and commuters who wanted to conserve agricultural land and the beauties of rural England, the authorities responsible for roads, public transport and water supplies, and the mining and quarrying industries; the Labour movement and the hunger marchers from Jarrow and Clydeside; the educationalists, the tourist trade, and even the Ministry of Defence.
>
> (Donnison and Soto, 1980, p. 5)

The diversity of these groups is evident from this list and it is not surprising that the consensus was fragile or that the purpose of planning was left vague.

Adjusting to growth

This section outlines the dominant changes in direction taken by planning in the period between its establishment in the 1940s and the Thatcher government of 1979.

The 1947 planning system contained radical proposals in state control over the right to develop land and an imposition of a 100 per cent betterment tax (for details see Cox, 1984; Cullingworth, 1980). As the National Federation of Property Owners said at the time this 'would mean, in effect, the nationalisation of a part of the owner's interest in the land, and thus strike at the very root of the principle of private enterprise in property' (quoted in Ambrose, 1986, p. 47). How were such radical proposals possible? Backwell and Dickens have suggested that one reason was the weak position of the property developers and supporting financial interests in war-time. Added to this was the antagonism of the general public to anyone profiting from the results of war damage. There was also the need for a major physical reconstruction job to replace war-damaged areas and provide sufficient housing against the background of limited finance and materials. It is not surprising that this job was viewed as being public-sector led and that the 1947 Act was framed in this light. However, in time, as economic prosperity developed, the property sector regained its strength and profits were to be made in developing city centres and providing for the needs of increasing affluence, particularly in the form of more housing. There was therefore pressure to remove the financial aspects of the planning system, and this occurred in the 1950s. Local authorities became increasingly dependent on developers for funds (Backwell and Dickens, 1978).

The new prosperity created further conflicts in the development of towns as the commercial centres underwent rebuilding. New roads were built to meet the demands of increased car ownership and new housing areas expanded on to agriculture land on the fringes of the built-up areas. Once again therefore planning was seen to have a role in mediating between these conflicting demands. However, in the new consensus of the 1960s, planning was to have a very different relationship to the market. Now the key approach was to be one of partnership with the private sector. The ministerial guidelines, *Town Centres; Approach to Renewal,* issued by central government to inform local authorities on how to approach city centre development, is a good example of this partnership approach (MHLG, 1962). It was suggested that the speed of development required a more flexible approach in which the private sector played a major role:

> Renewal cannot be undertaken without public support and it cannot be carried through without private enterprise. There is increasing evidence of readiness by private developers to collaborate with local authorities in this field and it is the Minister's policy to encourage this. In many towns today the initiative in redevelopment is coming from private developers and the local authority has to move fast to keep pace with them.
>
> (MHLG, 1962, p. 6)

In this partnership the role of planning was to aid in the process of land assembly and assist co-ordination. As the financial aspects of the system had been removed, planning had to operate through its regulatory powers and its negotiation with the private sector. The approach to development plans of the public-sector-oriented 1947 system was seen to be too cumbersome and inappropriate for the new private-sector-led circumstances. This led to a review of the development plan system in the 1960s and a more broad brush, flexible approach.

Reade (1987) defines the planners' role as corporatist, and critically analyses their relationship to developers. He sees the arrangement as having considerable advantages for both sides. Planners gain some status and a role in society and developers gain because they can get on with job of making profits behind a shield of apparent regulation and control which gives them legitimacy. He sees this partnership as resulting in a minimum of political interference because politicians will also be happy to allow sensitive issues to be taken out of the politics into this technical, corporatist arena (see also Ambrose, 1986). This arena is one of secrecy and hence there is no public entry into the process. The arrangement enables it to *appear* that the public interest is being served without any involvement or monitoring of this. Such an approach reinforces the technical, apolitical attitude to planning that was epitomised in the 1960s by planners' attachment to the 'systems' approach and corporate management.

However, the new consensus on the role of planning was itself challenged in the late 1960s and early 1970s as the difficulties in the economy grew and conflicting demands could not be satisfied. Thus there was considerable opposition to the actions of developers and the neglect of communities (see for example, Ambrose and Colenutt, 1975; Counter Information Services, 1973). In particular the apolitical, elitist and technocratic nature of planning was questioned. Donnison and Soto (1980) explain this change in terms of the general climate of reduced deference to authority, plus the expansion of the effects of planning into larger areas of cities, causing greater conflicts and public opposition. In the light of the numerous battles over development schemes at this time they conclude that the cosy elitist approach was at an end.

As a result of such pressures planning became more concerned with the social aspects of plans and sought an expression of people's needs through participation exercises. In more recent times this social strand has continued through the involvement of planners in such issues as inner city problems and the needs of women, ethnic minorities and those with disabilities. A reflection of these trends can be seen in the production of a strategic plan for London. The production of the first Greater London Development Plan in the late 1960s sought to shape the social and economic development of the capital but was rejected by the Layfield Panel of Inquiry for being too ambitious. However, in the revision of the Plan in the 1980s an even more ambitious attempt was made to involve the public and formulate a social dimension to the spatial policies.

Resolution of conflicts and shifting consensus

These trends in planning during the post-war period will now be related to the earlier discussion on the purposes of planning. In this section the idea of a 'shifting consensus' is adopted from Reade. He suggests that over the period, based upon the balance of conflicting forces, different conceptions of the legitimate scope and purpose of planning were defined. Each time this new definition formed the basis of a new consensus (Reade, 1987, p. 32).

The consensus of 1940, as indicated above, was built upon an alliance of interests which masked a variety of conflicts. Hebbert (1977) quotes Hayek, who points out that such a consensus can be achieved only through concealing the different aims it incorporates. It was possible to combine the appearance of consensus with the different aims only through an ambiguous and vague statement of the purpose of planning. It can be seen therefore that any fundamental discussion of such purpose was avoided. The most commonly expressed reason given was that planning would balance conflicting interests. Thus in introducing the 1947 Act to Parliament Silkin states that the objectives of planning are 'to secure a proper balance between the competing demands for land, so that all the land of the country is used in the best interests of the whole people' (quoted in Donnison and Soto, 1980, p. 4). However, ambiguity reigns as no indication is given of the criteria for determining the balance or defining the public interest. Glass (1959) expresses the relationship between establishing a consensus for planning and its vagueness of purpose in the following way:

> Town planning was advocated as a device for getting the best of all worlds: individualism and socialism; town and country; past and future; preservation and change. In other words, the planners promised the people that they could have their cake and eat it.
>
> This is an attractive doctrine: it presents so many different faces that it hardly seems to require scrutiny. It appeals to conservatives and socialists alike. This streak of ambivalence has certainly helped the British planning system in winning consent.
>
> (reproduced in Faludi, 1973, p. 58)

She also refers to the climate of the post-war period being conducive to the notion of 'public interest' and the effect this had on broadening the view of the purpose of planning. 'Planning was no longer mainly thought of as a code of regulations to repair the damage of *laissez-faire* and to prevent further damage in odd patches, but as an instrument of the Welfare State' (reproduced in Faludi, 1973, p. 50).

Foley (1960) shows how this vagueness continued through the 1950s. He also refers to the function of ambiguity in winning allegiance from a broad cross-section of politics and society. Within this ambiguous approach he identifies three different strands to planning ideology which often conflict with each other. The first strand is that expressed above by Silkin, the reconciling of competing claims for the use of land so as to create a balanced and orderly arrangement.

The second strand is the creation of a good physical environment and the third is to contribute to a social programme and a better community life. These different strands would obviously appeal in different ways to the various groups mentioned above by Glass. A particular problem of combining these different aims, discussed by Foley, is the co-existence of physical and social goals. The first of these can be discussed in a technical way which matches the image of apolitical professionalism adopted by the planner at the time. However, the social goals do not rest so comfortably within this approach. This is a theme taken up in considerable detail by Reade (1987). His message is that this technocratic approach of planners welded to the corporatism mentioned above obliterated any consideration of the social purposes of planning.

The 1960s in particular can be seen as a time when the technical approach of planners was paramount. The 'systems' view of planners as the helmsmen of the city (ship) guiding it along its charted course (McLoughlin, 1969, p. 86) is typical of the self-confident and apolitical stance of the period. Such a position is founded on assumptions of shared values in society and allows planners to continue to regard their actions as in the 'public interest' (Cooke, 1983). However, by this time the immediate problems of post-war reconstruction had passed and with it the financial powers to control the development process. There is a lack of congruence therefore between the aims and purposes set out by the technocratic, confident planner of this period and the resumed dominance of the market in the development process.

This position was severely strained by the challenge to planning that arose in the more conflict-ridden atmosphere of the late 1960s and 1970s. This was a challenge to the apolitical nature of planning, its elitist approach, and its claims to be the guardian of the public interest. As a result of community action and advocacy movements, planning adapted to the new situation by adopting additional purposes. Features that were now stressed included the involvement of the public, and concern for those people in society who were suffering from the declining economy. This brought to the top of the agenda the social implications and redistributive effects of planning (Eversley, 1973). However, as the powers of planning were not changed and the market still dominated the development process the ability to fulfil these broader social objectives was lacking. This led to further disillusionment and claims that the purpose of planning was to mystify or cloak reality with the pretence of social concern.

The conflict contained in the apparent consensus has been a central feature of the discussion so far. It is worth developing this aspect a little further as there are different dimensions to the conflict (see

Healey, 1983, for a fuller discussion of this). Attention has been given to the range of different groups and movements that forged the loose alliance of the 1940s. This has also been expressed in terms of conflicting economic interests (Backwell and Dickens, 1978). Ball (1983) examines the conflicts between landowners and builders, and planning's relationship to them. In some of these accounts a pluralist view is adopted and sometimes a neo-marxist; however, in both cases some role is seen for the state, and hence planning, in reaching a compromise between the conflicting interests.

However, a much broader level of conflict, resulting from the contradictory role of the state, is also referred to by many writers (see for example, Hague, 1984; Ambrose, 1986; Ball, 1983). As mentioned above, this contradiction is seen to stem from the need for the state simultaneously to foster capital accumulation while also preserving social harmony. Drawing on Habermas, as many of these writers do, Ambrose summarises this contradiction as a need for the state 'to safeguard the existing mode of production and the continuance of capital accumulation by the minority in market-dominated conditions, while at the same time maintaining the necessary level of mass loyalty to the system' (1986, p. 28). In planning terms this can produce the confusion of purpose illustrated above. On the one hand planning will need to foster and stimulate the private sector, for example through information, infrastructure and enabling functions, while also contributing to social harmony by appearing to act in the public interest, for example through providing a legitimising ideology, incorporating social aims or carrying out participation. As Healey (1983) points out these two functions are supported by different ideological justifications. The first rests on the notion of freedom and choice while the second draws on an appreciation of the existence of some form of 'public interest'.

From this broad contradiction flows another problem relating to the conflicting approaches to decision-making adopted by the market and state intervention. The accumulation role of the state requires an acceptance and fostering of market processes while the social harmony function requires state intervention. This market/intervention contradiction is mentioned by a number of writers (for example, Broadbent, 1977; Reade, 1987; Ravetz, 1980). Broadbent describes the conflict in planners' activities as they co-operate with the private sector by smoothing the path to development while also aspiring to some socially optimum allocation of uses. Reade and Ravetz also refer to an inherent contradiction between a market system of development and attempts to plan in the 'public interest'. They discuss, in particular, the way in which the Uthwatt Committee and the 1947 Act tried to address this problem by restraining the market element in the contradiction. With the repeal of the financial provisions in the 1950s such resolution was lost. Since then planning has had to live with this basic dichotomy. The subsequent resolution has been through presuming a position of power that in reality does not exist. Making planning a technical and unintelligible activity helps to obscure the reality. Thus according to this interpretation, the problem is resolved by allowing the market to remain the dominant process and setting up a system that appears to intervene in the 'public interest' but does not achieve much. However, if such a situation satisfies people then the social harmony function is still maintained. The challenge to planning in the 1960s can be seen as a challenge to this method of achieving harmony and subsequent attempts at participation as a new way in which planning can contribute to the state's harmonising role.

Another approach to conflicts of planning purpose, which is of particular relevance to the topic of this book because of its focus on planning law, is that of McAuslan (1980). He identifies three kinds of ideology that are, or potentially could be, part of the underlying basis of planning law. Although he accepts that the presentation of these ideologies is somewhat simplified he nevertheless claims that the difference between the three ideologies is significant and that their co-existence generates conflict. The three ideologies, in abbreviated form, are the protection of private property, the advancement of the public interest and the promotion of public participation. In very loose and broad terms these ideologies can be linked to the discussion above by suggesting that the first links to the market processes, the second to the attempts of planners to promote the public interest through its administrative and bureaucratic role and the third to the challenge to the elitism of planning that developed in the late 1960s.

Having identified these three conflicting ideologies McAuslan demonstrates how they differ greatly in their influence over decision-making. The first, the protection of private property, has the longest history and is the most dominant. It is supported by common law and, on the whole, the legal profession. The fact that private property is so fundamental to British society means that its supporting ideology 'provides the outer boundaries of any public involvement in private property and market ... and ensures that co-existence between public interest and private property will always be on private property's terms' (McAuslan, 1980, p. 145). One dimension of this dominance is the way in which lawyers and 'the law' are widely regarded as neutral and above power struggles whereas such support for private property rights can be regarded as essentially political.

McAuslan sees the second ideology developing in the nineteenth century with the regulations over property rights brought about by the conditions of the urban working class. Such conditions and the campaigns mounted to address them led to the acceptance, in the public interest, of government interference with property rights. This development of the ideology of public interest confers power on administrators through their implementation of the particular laws that are passed. These administrators are accountable to Parliament. Since the nineteenth century there has been a considerable increase in the number of laws passed in accordance with this ideology including the planning legislation. This has been added to common law causing a certain degree of conflict. McAuslan describes how there has been a 'constant oscillation by the courts between their desire to reassert the rights of private property against the all-pervading bureaucracy and their sense of obligation to uphold the lawfully constituted authority of government' (1980, p. 5).

Notwithstanding the antagonism between the ideology of private property and that of the public interest, McAuslan stresses their common aspects. They share a concern for maintaining the status quo and defending this against any attack such as that based upon the third ideology of participation. This status quo he defines as 'the existing state of property relations in society, the existing capitalist system with its emphasis on private property and a functioning market for that property' (1980, p. 268). A governing elite exists incorporating both the first two ideologies. It is described by McAuslan in the following terms:

> The law and administration of planning is operated, explained, interpreted, manipulated and occasionally reformed by judges, senior legal practitioners, public servants both central and local, well established professions such as chartered surveyors, land agents and valuers, Ministers and leading local councillors, and planners.
>
> (McAuslan, 1980, p. 268)

The third ideology, based on participation, stands in opposition to these two co-existing and reinforcing ideologies. The participatory ideology claims the right for all to be involved in decision-making, not just those with property, and denies that public servants operating their own conception of the public interest can satisfy people's needs. McAuslan identifies that in the 1970s some planners were propounding views on the purpose of planning that fell within this third ideology. Their egalitarian sentiments, their support of genuine participation and their identification with the substantive needs of those not involved in the decision-making processes all lent theoretical allegiance to this third oppositional ideology. Thus there is confusion over the purpose of planning between those that operate as part of the governing elite and those that challenge this operation adopting a participatory ideology. In addition, within the elite itself, conflicts between property rights and intervention in the name of the public interest can develop and require resolution.

Conclusions

Three broad conclusions are drawn from this look at post-war planning history. First, the changes to planning during the period can be divided into different phases. Second, the changes in the approach of planning match changes taking place in the role of the state as a whole. Third, that, notwithstanding the changes, there are certain purposes attached to planning that can be identified as enduring throughout the period.

The first phase is the 1940s, when thinking was dominated by inter-war problems and the experiences of war. The main features affecting both attitudes to the state and to planning were: the changing socio-economic situation; the need for a morale boosting vision of society; and the resultant all-party and elitist approach to a solution. The capitalist economic system was under threat from its bad performance in the inter-war period and examples of totalitarian alternatives. Economic planning was developing as a solution to these threats and this involved an acceptance of a new interventionist role for the state. A comprehensive approach to physical planning was to develop within this context, linked to the new economic approach and the rescue of democracy. Industrial restructuring leading to a dominant position for large industry in the new developing industrial sectors was placing its demands on the planning system.

The war had generated a need for a vision of society that was worth fighting for and which was different from that of the conflict ridden inter-war period. This set the scene for the Welfare State which was linked to the new state-orchestrated economic approach. Planning was seen to play its part in this reconstructed society and was propagated during the war as as an important part of this new vision. The proposed solution to the longer term social and economic needs, and the more immediate war-generated demands, was formulated in an elitist fashion by a small number of politicians, civil servants and experts. This applied to all aspects of government as well as town and country planning.

The consensus that evolved from this period incorporated a large number of different interests held together by common agreement over the need for a new approach and by the urgency of post-war social and physical reconstruction. This led to the necessity for a vagueness of planning purpose in order to satisfy everyone. When any purpose was expressed it was usually stated in terms of ensuring a balance between competing demands in order to safeguard the 'public interest'. Such an approach attracted popular support and enthusiasm because of planning's part in the new society to be created by the Welfare State.

The second phase runs from the 1950s to the late 1960s. The affluence of the 1950s meant that the extreme conditions of the depression and of war-time were less significant in influencing popular attitudes. A reaction to the constraints and policies of the immediate post-war years set in. As far as planning was concerned, with increasing profits to be made from development and the strengthening of property interests, the radical financial aspects of the 1947 Act were repealed. Control over the land market was reduced and development became private sector led. Government's major preoccupation was to maintain economic growth as the basis of the Keynsian/Welfare State approach and it pursued a corporatist means to achieve this. This elitist and co-operative relationship between the private sector and the state was reflected in planning. A partnership, negotiating approach was adopted and the planning system made more flexible. There was a heavy emphasis on the merits of technocracy in planning and government generally. Planning was viewed as a technical, apolitical activity based upon assumptions of a consensus of values in society. Thus planning was still viewed as the pursuit of the 'public interest' and it was carried out in an elitist fashion but lacked the means to fulfil its aims except through negotiation with the private sector.

The third phase is therefore from the late 1960s to 1979. The main characteristics of this period were the growing economic problems leading to the fiscal crisis of the state and the disintegration of the appearance of social coherence. Thus the period is epitomised by governments struggling to deal with economic problems, finding it difficult to maintain a growing economy sufficient to match public expenditure requirements and an increasingly dissatisfied population. The elitist, corporatist aspect of government was particularly challenged with demands for more participation. Planning as part of this elitist approach was similarly under attack and its ability to operate in the 'public interest' questioned. There were moves to open up planning with more participation and to consider the different interests in society rather than assume a common interest. An atmosphere of uncertainty set in as these socially oriented purposes of planning were given greater prominence but the planning system remained constrained and, to a great extent, elitist. While there was growth in the economy, planning could perceive its role in terms of steering and balancing the resulting demands and also seeking to achieve certain social ends through the redistribution of the resources generated by this growth. However, as economic difficulties grew, it became impossible to achieve social objectives without getting involved in priorities between groups or areas. This last aspect undermined the technocratic assumptions as it required taking sides and 'confronting the horrible possibility of having to enter the political arena' (Ravetz, 1980, p. 208).

Finally those aspects of planning purpose which were evident throughout the whole post-war period are summarised.

1 First, apart from maybe the immediate post-war years, the private sector dominated the development process. This is expressed in the limited powers of planning, restricted to control and negotiation, and the resultant dominance which the private sector has in any corporate style discussions. This dominant role is backed up by the overriding importance given in law to the ideology of private property rights.
2 However, throughout the period acknowledgement is given to the need for some kind of planning intervention based upon the concept of 'public interest' and balancing of interests. This acknowledgement has its roots in the Public Health legislation and was given a boost in the context of the Welfare State.
3 The elitist, corporatist approach to planning dominated the period. This had its origins in the expert/politician constructed package of the 1940s and continued as planning became part of the bureaucratic, administrative machinery of corporatism. This approach permitted a minimum of public involvement. From the late 1960s onwards this was challenged but with little effect on the overall planning system. The challenge was largely confined to aspirations and a few particular battles.
4 Throughout the period planning was fraught with crisis and uncertainty as it tried to deal with the basic conflicts and contradictions of the state itself. Planning played its part in stimulating growth, supporting the minority interests that dominated the development industry, easing the operation of the property market, for example in providing infrastructure or contributing to city centre redevelopment, and accepted the limits of a legal system based on property rights. On the other hand planning also contributed to the social harmony role of the state, contributing to the new post-war vision of society, and providing an appearance of intervention in the 'public interest'. The adoption of social objectives and participation in the 1970s can be seen as an extension of this function.
5 The attempts to incorporate a wide range of interests into the consensus plus the problem of dealing with the basic

contradictions of the state's role meant that throughout the period the purpose of planning was left vague and ambiguous. For most of the time planning was portrayed as an apolitical process, carrying out the technical job of balancing particular interests for the overall good. As this was challenged in the developing economic crisis so planning gave more emphasis to those social purposes that related to the more conflictual, participatory, society. Some of these socially oriented objectives had been fairly dormant since the 1940s when there had been a similar need to create a programme that could be seen as oriented directly towards people's needs. In essence then, planning during the period adopted, most, if not all, of the possible purposes outlined earlier. It gave emphasis to some of these purposes at particular times, but overall, conveyed a vague and unquestioning stance. Some of the purposes can be seen to relate mostly to the function of aiding the market, others to the function of maintaining social cohesion and the legitimacy of the system.

Chapter three

The nature of Thatcherism

The advent of Mrs Thatcher's government in 1979 brought a new approach to the British political scene. There is, however, a debate on just how radical this approach has been. This chapter provides an analysis of the most important aspects of Thatcherism as a background to the rest of the book. First, it is necessary to unravel the precise nature of Thatcherism, describing the various elements from which it is constituted and the degree of coherence between them. The challenge Thatcherism presents to earlier Conservative ideas and the post-war consensus is also discussed. Any break with the post-war consensus that might be established is seen as crucial to the discussion because, as outlined in the last chapter, town planning as a state activity was established as part of this consensus. Thus this chapter explores three interrelated questions: What are the distinctive features of Thatcherism? Do these features add up to a coherent ideology? How and in what way can they be described as radical? The discussion of these questions will provide the first stage in constructing a conceptual framework within which to explore changes to the planning system.

The elements of Thatcherism

Introduction

Notwithstanding their differences of opinion, most commentators on the Thatcher government accept that its philosophy derives from a variety of sources. A typical description is that of Riddell (1983), who states that Thatcherism is 'influenced by different, and often conflicting, creeds ranging from monetarism, via moral authoritarianism, to an anglicized view of Gaullism' (1983, p. 11). Many writers identify two strands in these influences which can be broadly described as concerned with, on the one hand, how the economy should be organized, and on the other, the style and content of government. It is claimed that since 1979 Britain 'has been moving both towards a freer, more competitive, more open economy and towards a more repressive, more authoritarian state' (Gamble, 1984, p. 8). The two strands have been given a number of different labels, for example, social market economy and authoritarian populism (Gamble, 1984), economic liberalism and authoritarianism (Edgar, 1983), neo-liberalism and combative Toryism (Norton and Aughey, 1981) or simply liberalism and conservatism (King, 1987). Often the analysis of the strands has focused on the duality of 'free market – strong state' (see Gamble, 1979a, 1988; S. Hall 1983; Eccleshall, 1984, pp. 108–11; Levitas, 1986a; Heald, 1983) and it has been suggested that the distinctiveness of Thatcherism derives from the unique combination of these two strands. Some authors, notably Stuart Hall, have utilised the concept of authoritarian populism to describe the second strand. However, it has been claimed, for example by Jessop *et al.* (1984), that this concept of authoritarian populism is lacking in clarity. As will be described later the populist element does seem to draw on both authoritarian and free market values and so it is felt more useful to start the analysis of Thatcherism by considering three elements: economic liberalism, authoritarianism and populism. The question of the relationships between the three will be discussed subsequently.

Economic liberalism

It was the economic ideas of the Thatcher government that attracted most attention in the early days of its administration. This was primarily caused by the outspoken advocacy of monetarism. Monetarism can be described as an approach to government economic policy in which it is thought that the control of money supply, including bank credit and government borrowing and spending, will check inflation. However, although the label of monetarism is commonly used to describe the Thatcher government, particularly in the early years, the label does not offer much understanding of the distinctiveness of the government's approach – after all the previous Labour government also adopted monetarist policies. Even though a distinction can be drawn between the approaches of the two parties to monetarism, along the line of whether such policies are part of a bundle of government economic actions or whether they are seen as the *only* permissible ones (see for example, Gamble, 1979a, p. 11), one has to turn to the rest of the government's attitudes to find the special nature of Thatcherism. In what broader ideological context does their monetarism operate?

Keith Joseph (1976a) indicates in his aptly titled booklet that

Monetarism is Not Enough. He points to the pressing need to generate a strong climate for entrepreneurship and risk-taking. He fears that 'monetary contraction in a mixed economy strangles the private sector unless the state sector contracts with it and reduces its take from the national income' (1976a, p. 52). Here we have the basis for a set of parallel policies promoting the free market and cutting back on state economic interference and state expenditure. The strong climate for entrepreneurship advocated by Joseph can be compared to the views of supply-side economists. They see a need for the reduction of taxation and regulation in order to energise the market economy. As will be seen later, such a view leads to an emphasis on removing unnecessary bureaucratic constraints, such as planning regulations, and ensuring the right material conditions such as land and infrastructure.

Adam Smith is constantly evoked in claiming that there is no role for government in redistributing resources and intervening in the economy. The new approach is a direct challenge to the Keynesian one and governments are blamed for playing a major contributing part in Britain's economic decline. For example Nigel Lawson, in a speech to the Institute of Fiscal Studies in March 1981, said 'if neo-Keynsian demand management was the necessary condition of economic growth, we would all still be living in caves and wearing woad' (quoted in Riddell, 1983, p. 80). According to the new approach the answer is to release the natural forces of the market order and generate growth again through the pursuit of individual profit. With the removal of state interference and the creation of free market conditions, a more competitive and productive environment will result and allow the full pursuit of such natural instincts as individual initiative, the acceptance of inequality and adoption of self-help. From such principles an image of society and human nature can be constructed such as the one outlined by Joseph and Sumption:

> A society of autonomous individuals is the natural condition of mankind.... Men are so constituted that it is natural to them to pursue private rather than public ends. This is a simple matter of observations. The duty of governments is to accommodate themselves to this immutable fact about human nature. Their object (and one must assume the original purpose for which they were created) is merely to avoid the inconveniences which attend the uncontrolled pursuit by private individuals of private ends.
>
> (Joseph and Sumption, 1979, pp. 100–1)

Thatcherism cannot be described simply as a return to classical liberalism although some of its principles have been used for rhetorical purposes. Such concepts as 'freedom', 'choice', 'competition' and

'individual initiative' have been employed to back up policies that reduce the power of trade unions, state intervention in the economy,
state expenditure and state enterprises. It is intended that the burden
on the private sector of high labour costs and high taxation to cover
a seemingly ever-increasing state expenditure will be removed. Riddell suggests 'the Thatcherite commitment is to private enterprise as
opposed to the public sector, not to the operation of competitive
markets as such' (1983, p. 169).

O'Sullivan (1976) has pointed out that there are difficulties in
adopting these liberal ideas to stem the expansion of the modern
state. The individualistic, competitive, self-seeking values that
underpin such an approach have been considerably eroded in society
today, particularly since the Second World War. Indeed, as Gamble
(1981) has noted, these values in the past were ones that Conservatives opposed because economic liberalism, 'promoted
individualism, questioned the authority of established institutions
and encouraged selfishness and competition between individuals and
social classes. The Conservatives tended to think that preserving the
social order was more important than preserving a market order'
(1981, p. 146). So if Thatcherism is to use these liberal values to promote a reduction in state activity then it has to pursue a strategy of
changing people's attitudes. Some authors (for example Gamble,
1981, pp. 162–4, Bosanquet, 1983, pp. 14–18) have illustrated how
the values of the market order are being blocked by the processes of
social democracy central to post-war society. The Thatcherist view is
that the assertion of political and social rights and the search for
equality, stimulated and pursued by organizations such as the trade
unions and the Labour Party, lead almost automatically to increased
state intervention, high public spending and vast bureaucracies.

According to this view democracy can produce demands that are
too great to satisfy and so care has to be taken to prevent such
demands developing. The values that support a major role for the
state, such as those justifying the establishment of the post-war Welfare State, have to be challenged and changed. Similarly, the
influence of those institutions that might foster such values has to be
contained. Now the problem arises of how to retain the legitimacy of
the state in its new minimal clothing when the values of the self-seeking individualistic market order are those that are being propagated.
Why should the state command any loyalty or allegiance especially
under conditions of inequality and high unemployment? It has been
observed that 'in so far as unimpeded market forces tend to generate
inequality, poverty, resentment and hostility, government must pay
closer attention to the problem of political stability' (Eccleshall,
1984, p. 109).

Traditionally Conservatives have called upon the concept of One-Nation as propounded by Disraeli. Namely that it is one of the duties of those that rule to take account of the 'condition of the people'. This leads to 'enlightened' policy oriented towards the less well off and as a result creates loyalty and deference in the masses and hence stability. After the Second World War the Conservative Party under the guidance of people like Butler and Macmillan came to accept that the role of the state in providing welfare and ensuring a buoyant economy and full employment was acceptable and could fulfil this need for legitimacy and stability. Thus the Thatcherist approach to the state threatens to discard these attributes and hence potentially to generate social instability. As will be seen later this is one of the criticisms made by the so-called 'wets' in the party. This problem of liberalism is neatly summarised by Norton and Aughey: 'a truly liberal model can never be satisfactorily accommodated by Conservatism precisely because it lacks any conception of order and authority not dictated by individual reason' (1981, p. 41).

Authoritarianism

National unity and stability have to be maintained by other means. This leads to the other main strand of Thatcherism, that of authoritarianism. This authoritarian strand in Thatcherism has been influenced by the ideas of the 'Neo-Conservatives' in the USA, whose views are elaborated, for example, in the journal *Public Interest* (see Edgar, 1983, 1986). One of the editors of this journal, Kristol, has pointed out that the ideas of Hayek and Friedman, and those of their followers, replace the concept of a 'just society' with that of a 'free society' and in so doing lose the legitimising myth that holds society together. He also draws attention to how the idea of a 'free society' assumes that society is composed of virtuous individuals. Which he contends is not the case. The reality is a society full of 'self-seeking, self-indulgence and just plain selfishness' which is 'ultimately subversive of the social order' (quoted in Edgar, 1983, p. 21; see also O'Sullivan, 1976, pp. 145–7). So the message is that the free market must be accompanied by a 'moral authority of tradition, and some public support for this authority'. This is where the seemingly contradictory combination of free market/strong state is generated.

Within Conservative circles in this country there has also been a concern to ensure that the enthusiastic adoption of the principles of economic liberalism does not undermine authority. For example it has been said that

The urgent need today is for the state to regain control over 'the people', to reassert its authority, and it is useless to imagine that this will be helped by some libertarian mishmash drawn from the writings of Adam Smith, John Stuart Mill, and the warmed-up milk of 19th century liberalism.

(Worsthorne in Cowling, 1978, quoted in Edgar, 1983, p. 24)

In 1977 a group of academics, commentators and politicians formed a society, which they called the Salisbury Group, to promote this brand of conservatism. The editor of their journal, Roger Scruton, is one of the more prolific writers in the group and for him the principal enemy of conservatism is not socialism but liberalism, 'with all its attendant trappings of individual autonomy and the "natural" rights of man' (1980, p. 16). He sees the need for authority which is generally accepted by the whole population and which can be used as a basis for action against any disturbing elements in society. As he puts it, providing the 'power to command and coerce those who would otherwise reform or destroy' (1980, p. 25). Now such an authority, in order to gain popular acceptance and have the strength to act in this way when necessary, must draw upon people's innate feelings which according to Scruton are 'manifest in patriotism, in custom, in respect for law, in loyalty to a leader or monarch, and in the willing acceptance of the privileges of those to whom privilege is granted' (1980, p. 26). Such feelings crystallise around the attitude of allegiance which Scruton contrasts with the liberal idea of society made up of an aggregate of individuals acting in their own interest. According to him, 'the healthy state or nation must command the allegiance of its subjects. Patriotism of some kind – the individual's sense of his identity with the social order – is politically indispensable' (1980, p. 35). Scruton sees the state taking on the important role of establishing this allegiance and ensuring that it is maintained. This role, which contrasts with the liberal minimal state, means that the state is an essential and strong force in society. It is required 'to uphold all those practices and institutions – among which of course, the family is pre-eminent – through which the habits of allegiance are acquired'. Then, once having established this authority, a strong state is needed to 'enforce that power in the face of every influence that opposes it' (1980, p. 33).

These principles set out by Scruton can be seen to have influenced the way in which the Thatcher governments have operated. According to Aughey 'the view of the Salisbury Group is much closer to the views of Mrs Thatcher than the scribblings of the neo-liberals' (1984, p. 18). The ingredients of this form of allegiance such as the family, patriotism and nationalism are constantly promoted and the necessary strength to uphold the authority has been sought through the increased finance, personnel and scope devoted to the forces of law and order. The Falklands War, the stance taken against the miners' strike in 1985 and the attitudes towards the urban riots all illustrate this authoritarian strand of Thatcherism. Some commentators believe that they can detect an increase in the importance of this strand, for example, Edgar claims that 'by early 1983 the authoritarian right seemed to have achieved a kind of hegemony over the Conservative party, and to have elbowed the pure free-marketeers to one side' (1983, p. 23).

The other side of the authoritarian coin is an undermining of democracy. O'Sullivan has described how Conservatives have always had their doubts about the doctrine of popular sovereignty and the way this doctrine has become part of democratic ideology. Popular sovereignty conflicts with the need for authority such as that set out by Scruton. O'Sullivan points out how such ideas of democracy can create a situation in which

> a modern government may, without absurdity, defend any policy at all, no matter how inimical to law, liberty and the security of property it may be, by merely claiming that it acted on behalf of the people, or in fulfillment of some electoral mandate.
>
> (O'Sullivan, 1976, pp. 18–19).

Although Conservatives may have been uncomfortable with the principles underlying democracy, it is only recently that they have openly questioned them. Scruton has no qualms in doing this. He likens democracy to a contagious disease that has raged through society wreaking havoc wherever it goes, even penetrating institutions such as the universities, the professions and the Church. He concludes that 'the unity between state and society demands no democratic process; indeed, at present, democracy is nothing better than a menacing disease' (1980, p. 69). Although the Thatcher government has been more circumspect in expressing its views, its abolition of the Metropolitan Authorities and the GLC, its lack of consultation in the conduct of its own affairs, its wish to reduce participation and willingness to set up undemocratic bodies all reflect this lack of commitment to democracy.

Populism

Having set out the economic-liberal and authoritarian elements of Thatcherism it is now time to turn to the third element – populism. Certain principles contained in these first two elements, for example monetarism, withdrawal of welfare provision or reduction in democracy, are not in themselves principles that are likely to command a great deal of public support. This is particularly so given the long period in which the Welfare State and social democracy have been accepted as immutable. Therefore one of the tasks of Thatcherism has been to turn the tide of opinion and translate the principles of economic-liberalism and authoritarianism into popular messages. As Stuart Hall (1983, p. 23) puts it, 'Thatcherism discovered a powerful means of translating economic doctrine into the language of experience, moral imperative and common sense'. He goes on to describe how the Thatcherite populism contains both themes from neo-liberalism such as self-interest, competitive individualism and anti-statism and also authoritarian themes such as nation, family, duty, authority, standards and traditionalism (see also S.Hall, 1988). Bogdanor also refers to the varied sources of Thatcher's populism when he says, 'Thatcherism ... owes as much to Mrs Whitehouse and to the folk image of the thrifty shopkeeper as it does to the economic foundations laid down by Milton Friedman and Friedrich von Hayek' (1983, p. 180).

In promoting its message the Thatcher government has drawn upon campaigns and groups such as those of Mrs Whitehouse, the Festival of Light, the Society for the Protection of the Unborn Child, the Monday Club, the National Association for Freedom and the numerous middle-class pressure groups that arose in 1974/5. The element that is common to these disparate groups is an antagonism to the drift of society towards a more permissive code of conduct and the fear that this will lead to social anarchy and an undermining of moral values. Worsthorne describes the feelings behind this movement in the following terms:

> The spectre haunting most ordinary people is neither that of the authoritarian state nor of Big Brother, but of other ordinary people being allowed to run wild. What they are worried about is crime, violence, disorder in the schools, promiscuity, idleness, pornography, football hooliganism, vandalism and urban terrorism.
>
> (quoted in Edgar, 1983, p. 21)

A speaker at the 1969 Conservative Party Conference produced a similar list. In his view 'the real crimes are not homicide, assault or battery, but the "violence" of dope peddling, abortion, easy divorce, pornography and homosexuality, immigration, town planning *[sic]*, and euthanasia', and according to the speaker all these crimes have one source, which is 'violence against a parent's self-respect, committed by the Socialist state apparatus crushing self-help and a parent's natural desire to work, to save and to insure and to educate his children independently' (quoted in Gamble, 1974, p. 111).

So Thatcher tapped these feelings that were very prevalent in the grass roots of the party and extended the appeal to the population at large. In a situation of economic crisis people's fears are accentuated and Hall has pointed out how Thatcher has been able to draw upon people's dissatisfaction with their experiences of the Welfare State such as its remoteness, its complicated and alienating bureaucracy and its inadequacy in many fields. However, Hall and others adopting his analysis, also believe that Thatcher has gone further than merely expressing these fears and discontents and has used them as the basis of a 'manufactured' populism. A 'siege mentality is ideologically manufactured whereby good Britons are commanded to remain alert to threats of national subversion' (Eccleshall, 1984, p. 110). The Falklands War presents a strong example of the utilisation of the bogey of the external threat to foster nationalism and unity behind the government. In addition, the internal threat to the national moral fibre is also posed. This internal threat encompasses scroungers on the welfare, unions, students and black people and any elements that create a threat to law and order. More recently, Thatcher's comments on the 'internal war' against the miners fits into this scenario. As a result of this populist approach the Thatcher government has been able to draw support for its programmes of economic liberalism, less state intervention, and a more authoritarian form of government. Stuart Hall describes this achievement in the following way;

> 'Being British' became once again identified with the restoration of competition and profitability; with tight money and sound finance ('You can't pay yourself more than you earn!!!') – the national economy debated on the model of the household budget. The essence of the British people was identified with self-reliance and personal responsibility, as against the image of the over-taxed individual, enervated by Welfare State 'coddling', his or her moral fibre irrevocably sapped by 'state handouts'.
>
> (S. Hall, 1983, p. 29)

In addition, the seige mentality generated by the government and the fears people have of the consequences for them of economic decline have created an acceptance of the need for strong leadership and allowed the authoritarian strand of Thatcherism to develop.

As far as planning is concerned such populist presentation of arguments can be detected in the way Thatcher governments have tapped antagonistic attitudes held by the general public. These revolve around such sentiments as bureaucratic meddling, high-handedness and arrogance and are epitomised by, for example, interference in individual liberty through control of the colour of bricks or size of back extensions or the disregard of people's desires in the building of high-rise flats. Such attitudes, though not necessarily related to reality, are used in a more general attack on planning to support the basic strands of the ideology.

A coherent ideology?

The principal ingredients of Thatcherism have been set out. The next part of this chapter discusses whether these ingredients are a random collection of views, perhaps evolving through pragmatic response to immediate pressures, or whether they are connected in such a way as to make a coherent whole. In the account of the three elements certain relationships have already been mentioned. For example, the need for legitimacy that stems from the economic liberal view generates the climate for the adoption of an authoritarian approach. Also the establishment of support for a Thatcherist ideology has required the populist programme which has drawn on values from both economic liberal and authoritarian elements. However, the question of coherence needs to be explored further if Thatcherism is to be used as a meaningful concept. In particular the seemingly contradictory nature of the economic liberal and authoritarian strands requires investigation.

As a number of authors have pointed out (for example Aughey, 1984, p. 2; King, 1987, pp. 126–33), there appear to be two broad positions on the question of whether Thatcherism can be regarded as a coherent concept. There are those, such as Stuart Hall (1983, 1988), that claim a consistency between all the elements of Thatcherism and attempt to develop a theoretical understanding of the coherence. Then there are those such as Riddell (1983) who say that labelling Thatcherism as an ideology is an artificial exercise because the actual actions of the Thatcher government demonstrate the lack of any consistency. In addition to this broad division of opinion there are different views about the nature of the link between the various elements.

Those that claim a coherence for Thatcherism suggest that the Thatcher government has had a clear conception of what it has been trying to do and that this conception leads to a drawing together of the various strands of which Thatcherism is composed. This viewpoint therefore stresses the links between the strands and attempts to show that they are not a random collection but functionally related. The argument presented by these writers is that the government is making a radical switch away from the post-war consensus policies to those of a market order. Their economic liberal philosophy expresses their economic intent and their authoritarianism the necessary strong state action to support it. A coherent strategy is attributed to Thatcherism in which an attack is being made on the way the economy has operated since the Second World War. This attack involves the development of an alternative economic strategy and the removal of aspects of the old order such as a welfare role for the state and co-operation with trade unions.

The populist attitude of the government is seen as a consciously created programme to mobilise support for their radical shift. As already noted, Stuart Hall talks about the way in which Thatcherism has translated a rather high-minded and abstract economic doctrine into a 'philosophy' that can be understood by the general population. With the help of the popular press the Thatcher message is being propagated to create a new 'common sense'. Support for the new approach to government is ensured by linking it to popular fears and sentiments. As Hall says 'it addresses real problems, real and lived experiences, real contradictions – and yet is able to represent them systematically into line with policies and class strategies of the right' (1983, p. 39). As mentioned above, this populist aspect of Thatcherism is regarded by these writers as a consciously constructed programme (indicated by the use of the word 'systematically' by Hall in the above quote). The aim of this populist programme is to generate the necessary support to stay in government and to ensure that the radical economic strategy which involves hardships such as redundancies and unemployment will nevertheless be accepted. For example, Bleaney claims that the experience of Heath's conflict with the miners and subsequent loss of the electoral argument 'has been a major formative influence on the whole Thatcherist programme. It has been the main impetus behind the notion that the problems of British society are at bottom ideological, and have first to be attacked on that level' (Bleaney, 1983, p. 139).

Thus, these writers seek to show that there is an identifiable ideology of Thatcherism; the various elements are linked, consciously created, and form an overall strategy, there is a clear desire for radical change and in the populist programme a process for mobilising support for that change. These writers stress the importance of considering together the ideas and practice of Thatcherism. For example, Gamble (1984) has stated that although it is important to distinguish conceptually between Thatcherist ideology and Thatcherist practice there is a close relationship between the two. Hall and Gamble explicitly set their views against those who consider such ideology as mere rhetoric with no impact on the 'real' world of economic forces.

One criticism of this 'coherence' position has come from Jessop *et al.* (1984, 1988). As already noted these authors object to the conflating of the two concepts of authoritarianism and populism (1984, p. 33). They also think that Hall's perspective places too much emphasis on ideology to the detriment of the political and institutional context. If these other aspects are given their due attention they believe that the ideological coherence disappears leaving an 'alliance of disparate forces around a self-contradictory programme' (1984, p. 38). One result of this ideological emphasis is a neglect of the economic and social interests that provide the supportive underpinning for Thatcherist programmes (see also Ross, 1983). The position of Hall is opposed for both neglecting the different economic and social interests contained in the so-called 'post-war consensus' and also the cleavages contained within Thatcherism. These different interests generate not only contradictions within Thatcherism but also a greater degree of continuity with the past than would appear from the emphasis on the ideological dimension of Thatcherism (see King, 1987, pp. 128–33, for a fuller account of this debate).

In their conclusion Jessop *et al.* do not actually reject the ideological perspective of Hall but request an extension of the analysis to cover the other dimensions that they outline. This would also seem to be acceptable to Hall who, in his comment on the Jessop *et al.* critique (S. Hall, 1985), also talks about Thatcherism as a 'multi-faceted phenomenon'. Levitas (1986a, p. 18) has pointed out that the difference of opinion and emphasis is due to different focuses of interest. Hall and others who emphasise the ideological aspect are concerned with the way in which Thatcherism has provided the government with necessary support as a context for their policies rather than with an explanation of the policies themselves. Those who are concerned primarily with explaining the implementation of policies will need to take a wider perspective. In her own work Levitas is concerned with exploring the ideological contradictions between the minimum view of the state adopted by the neo-liberals and the maximum view of the authoritarian strand. The Thatcherist view of the state will be explored in the next chapter but a few points raised by Levitas will be mentioned here in relation to the question of coherence between the two main strands.

She states that, on the face of it, the two strands are ideologically opposed, being based upon two very different concepts of human nature and the ideal society. The neo-liberal strand is based upon the concepts of accountability, efficiency and freedom. These concepts are interpreted in a particular way such that accountability is to those that pay, efficiency relates to demand rather than need and freedom refers to deregulation. However, these concepts seem to contradict the concepts of the conservative strand which are authority, allegiance and tradition. It has already been noted how antagonistic Scruton is towards liberal concepts of freedom. How can these differences be reconciled? Gamble and Hall have already indicated how the neo-liberal ideology requires a significant role for the state in order to create the necessary conditions for the market, hence the label, 'free economy – strong state'. Levitas (1986a) and Belsey (1986) have also pointed to the need for the neo-liberal to accept a role for the state. Hence the position of the two strands merges on certain issues. They highlight the way in which both strands accept that the state should ensure national security and law and order. This fits naturally into the authoritarian perspective but is also a requirement for the operations of the market. One of the consequences of this requirement is the concentration of power in the central state 'in order to establish and maintain the "deregulated" market' (Levitas, 1986a, p. 103). The neo-liberal writers also place considerable emphasis on legal processes as opposed to political processes as a better context for the operation of the market. This echoes the concerns of the authoritarian strand to downgrade democratic procedures. Belsey summarises these overlaps in the following way:

> The neo-liberals want a strong system of law to protect the market, and do not object to authoritarian measures to enforce it. In spite of constant appeals to the naturalness or spontaneity of the market system of capitalism, its order has to be enforced. The neo-Conservatives do not believe in the inevitability of the market, but find its harsh discipline a politically useful means of imposing authority.
>
> (Belsey, 1986, p. 193)

King (1987) has also investigated the relationship between the two strands of Thatcherism. His view is that the dominant strand is the neo-liberal one but that this cannot stand on its own as it has no theoretical ability to cope with the necessary role of the state. King also sees the neo-Conservative strand coming to the aid of the neo-liberals and providing the justification for a strong state. However, he points out that this strong state is of a particular kind. It does not result in the acceptance of public sector activity in general and in fact state action at the local level is severely constrained. The 'strong state' is therefore interpreted as action by central government. In his exploration of further overlaps between the two strands King emphasises the point that both strands agree on the limitation of 'social citizen rights'. These 'rights' refer to economic and welfare expectations, such as economic security, education or health provision, which have become embodied in the 'Welfare State'. The implementation of such rights tends to reduce inequalities and results in a more egalitarian society. The attempt by the state to achieve such ends runs counter to both the neo-liberal conceptions of the market order and the neo-conservative view of a natural heirarchical structure to society.

At the other end of the scale to Hall and the 'systematics' are those who stress pragmatism and the long-standing Conservative tradition of rejecting theory. It is said that Thatcherism is being endowed 'with a quality of intellectual consistency that can hardly have been observed in practice' (Rutherford, 1983). This view is developed by Riddell, who believes that 'Thatcherism is essentially an instinct, a series of moral values, and an approach to leadership rather than an ideology' (1983, p. 7). He seeks to demonstrate that in practice there is no evidence of any consistent approach. For example, the Thatcher administration did not pursue a free-market or monetarist programme when it countered market forces to hold down interest rates and gas prices and provide aid to British Steel. Public expenditure has also continued to rise. Mrs Thatcher has proved amenable to adapting to 'political realities' and has responded more to 'the failure of earlier policies and to short-term pressures than the implementation of a carefully worked out blueprint' (Riddell, 1983, p. 19). He points to the discrepancies between the 1979 election manifesto and the practice over the following four years and also to the rejection of the more radical ideas such as those in the alleged 'secret manifestos' or those considered by the Family Policy Group of senior ministers.

Although Riddell claims that in practice Thatcherism shows no consistency and is pursuing an almost Fabian gradualism, he does admit that its actions might add up to major changes over time. He accepts that when Mrs Thatcher became leader in 1975, a radical new approach began: 'she had a distinctive set of political values which were consciously different from the post-war consensus of state intervention and collective provision' (1983, p. 21). He also agrees with Gamble that the undermining of the values and institutions that support social democracy, for example emasculation of Labour-controlled councils, the privatisation of nationalised industries, reforming of trade unions and sale of council houses, could provide the basis for a more radical change in the longer term.

According to Riddell the spokesmen and supporters of the present government have overemphasised the coherence of its approach in claiming a radical new direction for British politics. At the same time Riddell does acknowledge that a significant change has taken place and that 'all the old certainties about, and expectations of, political behaviour and the impact of high unemployment have been shaken' (1983, p. 1). All schools of thought seem to agree that Thatcherism has a strong and clear image of the changes it wants to bring about in British society. Thus Gamble believes that 'The Thatcherites have a vision of what they want the country to be like in the 1990s and beyond' (1984, p. 9), while Riddell says, 'Mrs Thatcher knows what type of society she wants ... indeed, a hallmark of the

Thatcher administration has been its thinking about strategy' (1983, p. 232).

This wish to change society is very evident in some of the statements made by government spokesmen. For example Lawson describes the three years since 1979 as 'turning the tide' and the 'rebirth of Britain' (quoted in Riddell, 1983, p. 3). Mrs Thatcher herself has often talked of mounting a 'crusade' while Keith Joseph has perhaps made the most clear statements of how a new vision of society is needed to replace socialism. For example, in a speech in 1974 he pointed to past failures to do this when he said, 'the reality is that for thirty years Conservative governments did not consider it practicable to reverse the vast bulk of accumulating detritus of socialism' (quoted in Russel, 1978, p. 16).

So the change proposed is a destruction of the consensus or 'settlement' that followed the Second World War and a reversal of the directions set by Keynes and Beveridge, shared by both political parties and seemingly the population at large, and which led to the Welfare State. The first aim of Thatcher and her supporters was to change this whole climate of opinion, which has led some commentators to describe her style as 'thinking the unthinkable'. In order to undertake the task of changing opinion, 'the articulation of firm intellectual and principled foundations for Conservative policy was a felt need by Mrs Thatcher and those that shared her views' (Aughey, 1984, p. 3). This led, amongst other things, to the setting up of the Centre of Policy Studies with the stated long-term task of changing this climate of opinion. This aspect will be explored in more detail in the next chapter.

The threads of the discussion on whether Thatcherism is ideologically coherent will now be drawn together. At the outset two positions were described, one claiming that Thatcherism was coherent and the other disputing the fact. However, as the discussion has developed it has been shown that there is more common ground than might at first appear. The major area of disagreement centres on the issue of whether the actual practice of the Thatcher government matches its rhetorical statements. According to Gamble the general philosophical approach of Thatcherism does guide and interact with practice but he admits that at times there may be blockages and limitations in some areas of implementation. Riddell looks at the implementation and sees considerable departure from the rhetoric and therefore denies the existence of an ideology of Thatcherism. However, he does accept that many of the actions of the Thatcher government can be seen as setting the context for a longer term radical transformation of society. All the writers agree that Thatcherism has a clearly stated desire to change society and a radical vision of the future. There is also a consensus of opinion that Thatcherism has mounted a programme of mobilisation to gain support for the changes it proposes and that this mobilisation has met with a considerable degree of success.

Is Thatcherism Conservative?

This stance of Thatcherism in pursuing a radical programme of change stands in stark contrast to the generally propagated view of conservatism and also, seemingly, in contrast to post-war Conservative practice. In trying to understand the implications of Thatcherism it is important to know exactly how far the approach differs from conservatism of the past.

Thatcherism involves a conviction to change society but it is the essence of conservatism to oppose radical change. Does this mean that Thatcherism is not Conservative? Ian Gilmour (1977) has clearly expressed the reservations that he, and many others in the party, have about the dogmatic nature of the Conservative Party under Thatcher and its espousal of a radical re-orientation of society. He is very fond of quoting Disraeli in claiming that, while accepting some degree of change, British conservatism is not a system of ideas or ideology. He says;

> British Conservatism then is not an '-ism'. It is not an idea. Still less is it a system of ideas. It cannot be formulated in a series of propositions, which can be aggregated into a creed. It is not an ideology or doctrine, but at the same time, Conservatism is not opposed to change as such.
>
> (Gilmour, 1977, p. 121)

However, in his view the amount of change that is contemplated must be balanced by the overriding need to preserve stability. He suggests that the pursuit of dogma or ideology can disturb this stability and that 'the acceptance of some unpalatable measures is more likely to lead to stability than is continued and relentless struggle' (1977, p. 123). He even goes as far as suggesting that one of these unpalatable elements that has to be accepted is the winning of an election by the Labour Party. Of course, Gilmour is not the only Conservative to seek to temper change in the interest of stability; a constant theme of conservative thought has been the concern for 'small, manageable, piecemeal change, for small changes are easily absorbed by society' (Norton and Aughey, 1981, p. 20).

The fear that the more extreme views of Thatcher could lead to instability lies behind the apprehension within Conservative circles. It has been suggested that the party will become identified as a class-based party and lose its identification with the 'nation'. Again drawing heavily on Disraeli and his 'One-Nation' theme, it is suggested that the party should be above sectoral interests and should concern itself with the 'condition of the people'. Such a position would lead to an acceptance of the welfare role of the state although not necessarily in its present form, for example Pym favours privatisation in some areas (1984, p. 125) and Gilmour talks about the need to 'prune the rose' (1977, p. 152). However, a more compassionate approach is called for which has led to criticisms of Thatcher, for example in her disregard for unemployment or neglect of the human side of the miners' case. The demand is for a more 'balanced approach', 'a middle way' or an acceptance of the 'politics of consent'. As Norton and Aughey (1981) identify, the question is whether society is sufficiently cohesive (or forced into cohesion?) to withstand the disturbance of the balance between the rights of property and wealth on the one hand and the duties and obligations to the community and nation on the other. Two problems are presented by Thatcher's shifting of the balance towards the former. First, as Gilmour (1977) points out, it implies that 'almost the whole Tory Party has been marching in the wrong direction for thirty years' and second, it raises the issue of whether too strong an adherence to a particular set of economic principles could lead to a neglect of certain needs within the community and threaten stability. These fears that dogma and a bias towards the well-off might lead to instability have long roots in the Conservative Party.

In their interesting classification Norton and Aughey identify six varieties of conservatism. These are not mutually exclusive and particular governments can draw upon more than one variety. The views of Gilmour and others mentioned above seem to draw on two of these varieties. First, there is the viewpoint that Norton and Aughey call 'Paternalistic Toryism'. This viewpoint sees a natural order in society with a small elite having the responsibility of taking decisions and governing. Drawing on Burke, it is believed that 'society needs rulers leisured and cultured enough and independent of money-grubbing to perceive and act upon the national interest' (quoted in Norton and Aughey, 1981, p. 73). The elite is expected to be enlightened and always show concern for the governed – this is seen as a duty and obligation. Such an attitude leads to social cohesion and stability and is based upon concepts of integrity of leaders and loyalty of the masses. The leaders who are best able to fulfil this role are not necessarily found in the ranks of the meritocracy or intelligentsia but in those who have traditionally fulfilled this role, namely the landed aristocracy. Leadership drawn from this source is most likely to secure loyalty and preserve continuity and stability. One of the characteristics of British conservatism has been the way in which it has retained and absorbed old values while adopting new positions that reflect changing circumstances. Thus it has retained its relevance while also presenting an appearance of continuity. Wiener (1981) has described how the development of industrialisation created a 'compromise culture' in which the culture of the old landed aristocracy was retained and provided respectability for a party that would otherwise be seen as purely a businessmen's party. However, tensions arise as the old culture tends to be backward looking and a potential drag on economic progress. Weiner suggests that Mrs Thatcher's biggest job may be to challenge this old culture. Perhaps with economic growth so difficult to achieve the luxury of retaining the old culture for continuity's sake is too big a price to pay.

The way in which the Paternalistic Tories have sought to enter the world of the twentieth century is through the adoption of the values expressed by the variety of conservatism that Norton and Aughey call 'Progressive Toryism'. The Progressive Tory will seek to adapt the philosophy of Paternalistic Toryism to meet the challenge of democracy and changing aspirations. As Norton and Aughey state, 'principles are vague enough, historical myths sufficient, to enable the tory progressive to select his philosophical mentors judiciously and to make out a good case for a changed emphasis in policy and for a change in strategy' (1981, p. 77). The best example of such a Progressive Tory rethink occurred after the 1945 election defeat. There followed a period of intellectual reinterpretation of principle, led by Butler, to meet the changed circumstances of the post-war world. It is claimed as a result that Progressive Toryism 'is neither to the left or to the right but represents a judicious middle way between the extremes; that it represents a traditional Conservative concept of "balance", the golden mean between laissez-faire and an omnipotent State' (Norton and Aughey, 1981, p. 79).

The essence of this Progressive Tory 'middle way' is to preserve stability and unity through the consensus created by a 'One-Nation' approach and also to accept that the state has an important role in contributing to this. In Ian Gilmour's much quoted phrase, this will prevent a 'retreat behind the privet hedge into a world of narrow class interests and selfish concerns'. Gilmour and other Progressive Tories are keen to defend the post-war consensus position of the Conservative Party which included an acceptance of the Welfare State. He says:

> the state had a duty to try to make life tolerable for the least well-off and to give everybody the chance to develop his ability. Britain in the Tory view, was a great nation, not a random collection of

individuals. In that nation everybody had rights and duties, and those who were especially well endowed with rights had a duty to make a free economy bearable to all.

(Gilmour, 1977, pp. 155–6)

In a similar vein, Francis Pym describes the progressive position as the 'politics of consent' which is based on the idea that it is 'more effective and attractive if governments try and win the consent of the nation as a whole rather than railroad people into partisan decisions'. He goes on to say that 'such an approach calls for an understanding of people and of circumstances. It calls for a harmony between individual ambition and wider social considerations' (1984, p.x).

Given the strong conviction of the Thatcher government to break with the post-war consensus, and the strong association of Progressive Tories with that consensus, it is not difficult to understand the antagonism between the Thatcherites and the progressives or 'wets' as they have been called. The principal issue at stake in this antagonism concerns the role of the state. According to Russel, the dispute within the party is based upon the 'party's attitude towards the degree of government involvement in economic and industrial affairs, the size of the state or public sector, the level of government expenditure and even the future shape of the Welfare State' (1978, p. 13). He points to the 1976 strategy document *The Right Approach* for an acknowledgement of this dispute. In this document it states, 'The precise limits that should be placed on intervention by the State are reasonably the subject of debate within the party'. Behrens (1980) develops this further and describes divisions in Conservative politics after 1974 as a distinction between 'Diehards' and 'Ditchers'. Diehards believe that since the Second World War Conservatives have lost their fundamental principles and become collectivists. They need to regain their concern for limited government, individual freedom and moral rectitude. Ditchers, on the other hand, claim that policies must be changed in the light of circumstances. Although each category may include a wide range of people and groups, Behrens claims that there is an important distinction between these two categories. This distinction concerns attitudes to the state. Thus he says, 'what delineates the Diehards is their basic assumption that the post-war settlement had failed because of the over-weening interference of the state in society', whereas the 'Ditchers were united in their belief that post-war interventionism for all its faults, was based on the principles essential for a sound body politic' (Behrens, 1980, p. 8).

Although the labels attached to the factions vary there is general agreement that fairly fundamental differences of opinion exist within the Conservative Party. As described above these differences stem from alternative interpretations of how stability is to be maintained. The opponents of Thatcherism fear the disruptive effects of her radical, dogmatic stance and in particular the neglect of social consequences. A key issue of debate is the role that the state might play in looking after the 'condition of the people' and therefore in helping to promote stability. Having established some of the major differences between schools of Conservative thought, the next section will explore briefly whether there is any continuity between Thatcherism and post-war conservatism.

Thatcherism and continuity

At a pre-election strategy conference at the Selsdon Park Hotel in January 1970, the Conservative Party devised an approach that broke with the post-war consensus and was given the label 'competition policy'.

> It prepared to end the discredited interventionist strategy of the previous decade, to close down all the boards and agencies of intervention, to reintroduce criteria of competitive efficiency into industry, to undertake a big reduction in taxation, and to introduce legislation that would reduce the bargaining strength of the trade unions.
>
> (Gamble, 1974, p. 221).

The Heath government adopted this approach for two years before the conflict with the miners and a return to the previous corporatist policies. It has been suggested that Mrs Thatcher is merely readopting these ideas of 'competition policy' and pursuing them with more conviction: a resolute 'Selsdon Woman'.

Until the Selsdon 'competition policy' both parties in power during the 1960s had pursued what has been called a 'corporatist approach'. In this approach, as mentioned in the last chapter, one of the roles of the state is to provide a stable environment for capital accumulation. Thus the state is a creative factor ensuring the efficiency of the national economy. According to the advocates of this view the state needs to adopt this role because of the complex interdependent nature of modern society. It is an approach that has its origins back in the 1930s and brings to government the attributes of good industrial management to foster increased overall national economic growth. Heath was one of the breed of Conservative politicians following this 'modern' managerial approach. 'His style was managerial and middle class. Heath's strength was his single-minded concentration on policy and the details of effective administration of the state' (Gamble, 1974, p. 217). It has been suggested that his background and commitment to a 'corporate approach' meant that he was never fully convinced of the new 'competition policy' which had developed because of pressures from the right of the party and the perceived need to distinguish a Conservative programme from that of the Labour Party. According to Norton and Aughey, 'despite the rhetoric that was misguidedly taken as an indication of Heath's commitment to some resurrection of laissez-faire, the truth was that the Conservative leader was commited to increasing the effectiveness of co-operation between state and private sectors of the economy' (1981, p. 85).

These doubts over Heath's commitment present a contrast to Thatcher's conviction. For her there are no U-turns. It has been said that the policies of Heath's 'competition policy' lacked coherence or any overall theme, whereas Thatcher's approach is part of a well-worked-out ideology. Aughey (1984) contrasts the two leaders in the following way: Heath was a technocrat concerned with efficiency and effectiveness and if one set of policies did not seem to be working then he was quite happy to try out a different and hopefully better set, Thatcher on the other hand is an ideologue who will contemplate policies in the first place only if they embody certain principles and fit into an overall view of politics and society.

It is also suggested that Thatcher's conviction approach stems to some extent from a realisation that Heath had insufficient support for his change of direction when the crunch came. Those developing the Thatcherist approach were fully aware of this and they devoted much attention to the question of changing people's values and presenting a mass appeal – the populist strand described earlier.

> The right grasped what the modern wing did not, that a competition policy, however necessary, could not be recommended to the electorate and to the party on technical grounds alone, but had to become part of a much more general political and ideological offensive.
>
> (Gamble, 1974, p. 102).

Thus, although there are similarities in some of the policies pursued by Heath between 1970 and 1972 and those of Thatcherism, the broader political philosophy and programme surrounding these policies is very different. The Thatcher policies are backed up by a more coherent attitude to the desired direction of society as a whole and a concerted attack on past values. Heath was still imbued with attitudes of corporatism that meant he took a far less ruthless stance, for example on the Welfare State, and this allowed him to reverse his policies when the going was hard. In addition, he failed to mount a programme of popular support for his views. He lacked the populist and authoritarian attributes of Thatcherism.

However, some writers have suggested that, whatever the ideological differences between Thatcher and Heath, if one looks at the actions of the two governments considerable continuity can be observed. Riddell (1983), as mentioned earlier, highlights the continuation of interventionist policies such as protection of the privileges of the Stock Exchange and tax relief for owner occupiers. He also demonstrates the continuation of rising public expenditure – one of the main economic targets of Thatcherism. However, he also shows how this rising public expenditure incorporates important shifts and differences between expenditure heads. The largest decline in expenditure occurred in housing, while the increases were in defence, law and order, employment and social services. These changes to a considerable extent reflect the shift from a welfare to a strong authoritarian state. Riddell also points out that after the cutbacks under the previous Labour government, Thatcher had little scope for further large decreases in public expenditure without some radical reappraisal of the role of government. Although there have been many Think-Tank reports on how this can be done, Riddell sees no sign of the political will needed to put them into practice and he concluded, in 1983, that 'there has been no overt challenge to the post-Beveridge consensus' (1983, p. 139).

Finally, there are some central themes of conservatism that have been ignored up to now because they generate common agreement amongst the different factions within the Conservative Party. The first of these is support for the law and constitution. Gamble (1983) has noted that Conservatives have always seen their role as defending key institutions against radical reform and he sees 'no break of continuity in the use Thatcher has made of that (other) Conservative tradition, the defence of national institutions' (1983, p. 11).

A second area of agreement concerns the importance of strong leadership and a sense of patriotism. According to Behrens, strong leadership is advocated by Conservatives because of their belief in the 'unreason of man ... imbued with original sin' (1980, p. 15). There can be no doubt that Thatcher has presented the image of strong leadership and this was linked to patriotism in the Falklands War. Behrens has pointed out that this view of leadership is one area of continuity with Heath, although it is clear that Thatcher has put the principle into practice rather more effectively than Heath ever did. In addition, the principle is not one that causes any qualms for Conservative 'wets'. As mentioned above, they draw on the Paternalistic Tory view of an ordered society with strong enlightened leaders. One of the aspects of Mrs Thatcher that Francis Pym picks out for praise is the way she has 'returned decisive leadership to the forefront of British politics' (1984, p. 11).

The third area of agreement concerns the importance attached to the protection of the acquisition of wealth and the rights of private property. Norton and Aughey (1981) show how Conservatives justify this belief on moral, political and economic grounds. Moral because the possession of private property is an expression of human personality; political because it creates an area of life free from government interference; and economic because it is a productive way of running society. They go on to say 'ownership facilitates independence and encourages a sense of good husbandry and caution that is essential to the stability of the social and political order' (1981, p. 33). These views are obviously shared by Gilmour (1977) who links property to the family and presents them as the vehicles for protecting the individual and promoting freedom. He says, 'in their defence of the individual against socialism and excessive state power, Conservatives rely chiefly upon the family and private property. The family is the citadel of individual freedom, but that citadel needs its moat of private property' (1977, pp. 148–9). The widespread acceptance of these principles means that a policy such as the sale of council houses gains universal Conservative approval.

So there are certain areas of continuity between Thatcherism and past Conservative approaches. In particular, there is the re-affirmation of the importance of the rule of law and the constitution, the importance of leadership and the protection of private property. There seems to be a similarity in certain policy areas between actions since 1979 and those taken by Conservative administrations in the past, particularly between 1970 and 1972. However, this is overshadowed by the broader ideological context, the mobilisation of support and the coherent statement of future visions. Notwithstanding the similarities, this vision dominates most policy areas and generates new directions.

Conclusions

Thatcherism embodies an economic liberal strand which comprises monetarist policies, the facilitation of the free market and a desire to reduce state intervention, state enterprise and expenditure. The values that support this strand are antagonistic to those that have dominated society since the Second World War. A second strand is that of authoritarianism which involves a strengthening of patriotism and allegiance to the state and the machinery of law and order. This authoritarian strand is supportive of the economic liberal position.

This duality of free market/strong state is supplemented by a populism that translated the government's position into everyday terms and ensures support.

There are differences of opinion over the degree of coherence between the elements of Thatcherism although this centres less on the question of coherence between the various ideas and sources of inspiration than upon the degree of coherence between these ideas and practice. There is agreement that not all practice can be assumed to conform to the ideological guidelines and that each case has to be examined in detail.

This chapter has also looked at the question of whether Thatcherism is new and radical when compared to previous Conservative thought. Commentators seem to turn to botanical metaphors when considering this question. Norton and Aughey (1981), quoting Mendoza, describe the various dimensions that have historically made up the philosophy of conservatism as 'this tangle of old and new, these secular oaks, sturdy shrubs, beautiful parasitic creepers' (1981, p. 61). However, they refrain from saying whether Thatcherism is an oak or parasitic creeper! In a more recent paper Aughey does say that Thatcherism's 'roots lie deep in the party and it is no rare exotic growth' (1984 p. 3). Scruton (1980) does not seem to agree with this when he talks about a beautiful bloom dying at night. Meanwhile Gamble suggests that Thatcherism 'has meant the blooming of a thousand exotic ideological flowers and rank weeds where only a parched desert with the occasional cactus existed before' (1983, p. 10).

Thatcherism maintains certain themes and principles that have always been central to conservatism. On the other hand, major changes are also proposed that imply a radical shift from Conservative attitudes since the Second World War. The post-war consensus is directly challenged in a comprehensive way encompassing its value system and the institutional structure that underpins it. The radical change and the manner and consequences of the new strategy has provoked criticism within Conservatives' own ranks. There is considerable fear that the dogmatic approach and its lack of compassion could create instability.

The Thatcher government is not following a classic liberal line in its economic policy, although it utilises key words such as 'competition', 'free choice' or 'individual initiative' in its rhetoric. Riddell and Gamble, who are representative of the differing views on the coherence of Thatcherism, nevertheless agree that the net effect of the Thatcher economic policy is drastically to reduce the functions, services and employment provided by the state. However, the problem faced by Thatcher's government in 1979 was that Labour had already cut all the easier options. Further cuts required a challenge to the fundamental purpose of state intervention. Thatcherism is bringing into question the basic values that justify this state role, such as concepts of social justice, equality, or positive discrimination. However, this threatens the legitimacy of the state and the Thatcherist support for economic liberalism has only the concept of free competion to put in its place. An alternative form of legitimacy is therefore developed in terms of allegiance to a strong British identity and a strong purposeful government that is not afraid of taking difficult decisions to obtain the end results that will be in everyone's interest. King has shown how the antagonism to the values of the Welfare State is one of the key elements uniting the seemingly contradictory strands of Thatcherism.

Even though some writers, such as Riddell, are sceptical about the practical effect of Thatcherism there is general agreement that the arrival of Thatcher in 1979 set the scene for important changes. It is agreed that Thatcherism has made huge strides in changing attitudes and values that formed the basis of post-war policy and that Think-Tanks are 'thinking the even more unthinkable'. The longer a Conservative government following Thatcherist principles stays in power the more entrenched the new values are likely to become. Even Riddell commenting on the second term of office said that this would be 'more far reaching and would strike at the heart of the post-war settlement' (1983, p. 238). It is believed that once this ideological victory has been firmly established then the door will be open to more extensive and speedier policies. The third term has witnessed the further extension of the ideology into such key areas of policy as education, health and housing. According to Gamble, Thatcherism's 'real achievements have been ideological – shifting the focus of political debate, and making a chain of future institutional and legislative changes possible' (1984, p. 9). He suggests that the speed of change in different policy areas will vary according to how entrenched the social democratic values are in that particular area. This may explain why the Thatcher government did not fundamentally tackle health and education until its third term. Where does town planning stand in this respect and could the institutional and legislative changes occur rapidly and without much opposition? Planning has a long tradition of working alongside and in partnership with the market yet also has obligations to participate with the public and incorporate the community interest.

Predictions for the future go beyond suggesting a more extensive and speedier application of Thatcherite policies. It is also being said that the Thatcher crusade is not simply about retaining power but changing society (see for example, Gamble 1984, 1988; King 1987, p. 26; Jessop *et al.* 1984, p. 50). Thatcherism will outlive Thatcher.

'The Thatcherites are not interested in political ping-pong. They want policies and priorities established that are irreversible, which any future government is obliged to continue and dare not reverse' (Gamble, 1984, p. 9). A classic policy in this respect is said to be the sale of council houses. It is suggested that attention has to be given to what is happening in the workplace, office and school in terms of a changing climate of opinion. Changes in trade union legislation, education policy, sympathetic appointments in the media and Civil Service, denationalisation, privatisation, the assault on local government, could all contribute to the institutionalisation of policies based on New Right values. Is the 'Thatcherite consensus' even changing attitudes in the other political parties?

It has been the intention of this chapter to take a wide-ranging approach to the question of 'What is the nature of Thatcherism'. The broad level of the debate is somewhat removed from the details of planning practice and some means of establishing the linkage is now needed. Some of the aspects covered in this chapter will have greater influence on town planning than others. As pointed out earlier, town planning since the Second World War has predominantly been carried out by the state. Therefore one key aspect of Thatcherism that needs to be pursued further is its attitude to state activity. It has been shown in this chapter that the role of the state is a central question in Thatcherism; the reappraisal of state activity, the aim to reduce state expenditure and the desire to change the values and institutional structures of the post-war consensus. It has also been shown that the role of the state is a key issue in the conflict of opinion within Conservative circles. The next chapter will therefore look in more detail at this key issue: the role of the state under Thatcherism. In doing this an attempt will be made to provide a bridge between the broad debates about Thatcherism and town planning practice and to develop a conceptual framework for the rest of the book. Many of the debates outlined in this chapter can be satisfactorily developed only in relation to a detailed analysis of particular areas of practice. Once the particular area of planning legislation has been pursued in later chapters it will be possible to return to some of the issues raised here.

Chapter four

Re-orienting the state

Thatcherism has been built on a body of right-wing academic ideas. Particularly important to this study are those ideas concerning the re-orientation of the role of the state, as these would be expected to have a major impact on planning. Whether or not this has happened is the focus of later chapters. In this chapter the ideas of the most influential academics, particularly Friedman and Hayek, are explored. Their reasons for proposing less state intervention and their views on the acceptable residual functions of the state are outlined. The chapter then turns to the issue of how these ideas have been absorbed into the political arena, concentrating on the work of Keith Joseph and Mrs Thatcher who laid the ideological foundations for the 1979 election campaign and the subsequent 'rule' of Thatcherism. The main themes of the chapter are incorporated into a conceptual framework to guide the analysis in the rest of the book.

The academic foundations

Why less government?

Friedman has stated his basic philosophy in the following terms: 'reliance on the freedom of people to control their own lives in accordance with their own values is the surest way to achieve the full potential of a great society' (M. Friedman and R. Friedman, 1980, p. 359). According to him freedom is generated by the creative force of the free market, individual action and initiative. This creative force can be realised only if government interference is constrained. The best check on the encroaching power of government is to ensure that the political and economic spheres remain as separate as possible and that power is diffused throughout society by means of the free market. This prevents the build-up of political influence which can distort the functioning of society. Not only is the free market the best check on over-ambitious government activity and political bias, but also it is the best way of organising the economic rewards in society and creating the optimum degree of prosperity.

Hayek also believes that society has made a mistake in abandoning the principles of liberalism and he sets himself the intellectual task of demonstrating this point and drawing up the requirements of a good liberal society. Unlike Friedman he does not take liberty or freedom as an absolute value but believes that its usefulness to society has to be demonstrated. He seeks to show that only through liberty can society achieve progress. He believes that the rules that evolve through custom and tradition will in practice turn out to be those that allow a considerable amount of individual freedom. Any attempt to impose any other direction on society, such as socialism, will only put the clock back and be detrimental to progress. Hayek's view is that socialism may have been appropriate to the small hunting and gathering societies of the distant past but not to the complexities of modern society where socialist plans and prescriptions can only hinder progress. 'Socialism is simply a reassertion of that tribal ethics whose gradual weakening had made an approach to the Great Society possible' (Hayek, 1982, vol. 2, pp. 133–4). Through the historical evolution of individual liberty a vast creative force was unleashed that led to innovation, inspiration and the whole scientific and artistic development of society. Therefore liberty is to be pursued because this is the only way of maintaining progress. As Brittan has pointed out, for Hayek 'liberty is, in the last analysis, an instrumental value in the service not of happiness or welfare, but of progress – material and intellectual – seen as an end in itself' (Brittan, 1980, p. 34).

Although Hayek's aim in his well-known book *The Road to Serfdom* (1944) was to oppose socialism as it was developing in the 1940s, he sees his arguments being applicable to any kind of public intervention or 'collectivism', even if practised by Conservative politicians. This expansion of his philosophy is developed in his later work. There are a number of basic reasons why he sees socialism, collectivism and planning as unacceptable. One of these reasons concerns the complexity of modern society mentioned above. 'Society and economic processes are so complex that they are completely beyond the capability of any planner or planners to comprehend' (E. Butler, 1983, p. 71). There will be a limit to the amount of information that can be handled and therefore planners will resort to harmful simplicity. He considers that a much stronger foundation for dealing with complexity can be found in individual actions operating in the market and in rules that have slowly developed, through trial and error, over generations.

Friedman also applies his ideas to the operation of past governments.

He discusses the false trail that was adopted by governments since the Second World War, particularly in the USA and Britain. In his view the US government was responsible, through the Federal Reserve System, for the inter-war depression. However, popular opinion seemed to blame not government but the capitalist market. Thus according to Friedman this has led to the false faith in further government action and opened the door 'to men of good intentions and good will who wish to reform us' (1962, p. 201). Such faith led to the New Deal in the US and the Welfare State in Britain and these packages of action, and the increased role of the state that was involved, once initiated became imbued with an unwarranted permanency. According to Friedman the result has been more than forty years of overactive state interference. In his more recent work he reinforces this view and concludes that governments have increased their paternalistic role to evil proportions. 'Their major evil is their effect on the fabric of our society. They weaken the family; reduce the incentive to work, save and innovate; reduce the accumulation of capital; and limit our freedom' (1980, p. 158).

As mentioned above, in his later writing Hayek has switched his attack away from the 'pure' socialism of *The Road to Serfdom* to attack the more subtle approach of the Welfare State. He believes a principal reason for the growth of government intervention implied in the Welfare State is due to the confusion of two different kinds of law, that is 'the general rules of justice which enable a free society to grow and flourish without any central direction, and the organizational rules of authorities, aimed at achieving some particular social plan' (E. Butler, 1983, p. 121). The first kind of law is described as the 'rule of law'. It will not be 'made' but identified by judges from past experience. It will not be concerned with the interests of specific groups or the ambitions of public policy. The second type of law, which Hayek often calls 'legislation', is made by people; by Parliament. According to Hayek there is a need to re-establish the distinction between these two types of law and in so doing reduce arbitrary legislation and government activity. He sees that, with the disappearance of this distinction, it has become acceptable for the rules of society to be manipulated by human agencies who then seek to reform and redesign society instead of relying on universally agreed customs. Once the process has started there is no logical point at which to stop. 'To leave the law in the hands of elective governors is like leaving the cat in charge of the milk jug – there soon won't be any, at least no law in the sense in which it limits the discretionary power of government' (Hayek, 1982, vol. 3, p. 31).

As government intervention in the name of 'fairness' or 'social justice' has increased it has become necessary to abandon the universally applicable 'rule of law' and decisions are made which favour particular interests. Thus the 'rule of law' which has to apply equally to everyone is destroyed. It also means that concrete decisions have to be 'taken on their individual merits' which therefore requires delegation. The net result is that particular people are acquiring the authority to pass judgements. Not only that but also these judgements are necessarily arbitrary because of the vagueness of the concepts, such as 'social justice', on which they are based.

It is necessary to explore further the antagonism of Friedman and Hayek to the concepts and values of the Welfare State. This antagonism is a central aspect of their position and, as will be seen later, essential to the Thatcherite campaign. As discussed in Chapter Two, planners in the post-war period have operated in the context of the Welfare State and adopted its values, employing such concepts as 'the public interest' or 'community needs'.

Misconceived values

Hayek explores in considerable depth the vagueness of the values under-pinning state intervention. He points to aims such as 'common welfare' or 'social goals' and the emptiness of these concepts. Planners never seem to agree on the content or definition of their aims. According to him this is hardly surprising given the lack of commonly held and commonly prioritised set of values in society. Of course in the market system this can be avoided as everyone pursues their own ends. Hayek expresses the problem in the following way:

> The 'social goal', or 'common purpose', for which society is to be organized, is usually vaguely described as the 'common good' or the 'general welfare', or the 'general interest'. It does not need much reflection to see that these terms have no sufficiently definite meaning to determine a particular course of action. The welfare and the happiness of millions cannot be measured on a single scale of less and more. The welfare of a people, like the happiness of a man, depends on a great many things that can be provided in an infinite variety of combinations.... To direct all our activities according to a single plan presupposes that every one of our needs is given its rank in an order of values which must be complete enough to make it possible to decide between all the different courses between which the planner has to choose.
>
> (Hayek, 1944, pp. 42–3)

The above quote forms part of Hayek's attack on socialism in the 1940s but more recently, in the second volume of *Law, Legislation and Liberty*, he has developed and extended the argument to apply to the post-war expansion of government intervention. He sees that this expansion has been justified in the name of 'social justice'. Once again he aims to show that this is a meaningless concept and he pursues the argument with almost missionary zeal, 'I have come to feel strongly that the greatest service I can still render to my fellow men would be that I can make the speakers and writers among them thoroughly ashamed ever again to employ the term "social justice"' (1982, vol. 2, p. 97).

According to Hayek there are four reasons for the popularity of the term 'social justice'. In a dynamic market system there will always be some areas that are expanding and prospering while others are in decline – it is the very existence of this process that gives the market order its great strength and allows for progress. However, the impact of this process on those in the declining areas is immediate and visible while the benefits elsewhere are more hidden and diffuse. This means that those affected, instead of moving into the dynamic areas, can use their visibility to acquire power to support their claims in the name of 'social justice'. A second source of popularity is simply the expression of envy. This envy will be stated in terms of complaints that some people earn more than others. A third reason is that with the huge growth of bureaucracies, and large organisations generally, a significant proportion of the population is insulated from the workings of the market which they do not comprehend. Concepts such as 'social justice' could therefore be unthinkingly adopted in place of the principles of the market. The last reason Hayek proposes is that our instincts have not been properly overcome. As already mentioned, he believes that these instincts date back to tribal hunting and gathering societies where common aims and community roles could be established. However, in the complex modern society that has flourished under the market order such common agreements are impossible and unnecessary. Actions based upon concepts such as 'social justice' are therefore atavistic yearnings for a simple primitive past.

'Social justice' and similar concepts are used to imply that there is something wrong with the way wealth or other desirable goods are distributed in society and that some form of corrective action has to be taken in order to create a fairer situation. Hayek claims that such a view is meaningless in a free society. In a competitive economy it is the attributes of the individual such as skill, energy or initiative that determine rewards. In a free society there is no need to establish agreed common aims or purposes and instead everyone follows their own interests bound by a general system of rules which have evolved slowly over time and have obtained allegiance because of their proven usefulness. This 'rule of law' will become an almost unconsidered aspect of traditions and customs. It would be absolutely wrong to try and distort this pattern by imposing other values based on such concepts as 'social justice' which is, anyway, impossible to define.

The view that there is something wrong with the way the market distributes its benefits throughout society has led post-war governments to adopt actions to promote greater 'equality' of 'redistributive justice'. According to Friedman this has led to too much intervention. Friedman opposes equality as it conflicts with the overriding principle of freedom, through the taking away from some to give to the others. Concepts of 'fairness' or 'justice' for all lead to the problem of defining what is 'fair' and 'just' and who decides on the definition. He says:

> It has proved impossible to define 'fair shares' in a way that is generally acceptable, or to satisfy the members of the community that they are being treated 'fairly'. On the contrary, dissatisfaction has mounted with every additional attempt to implement equality of outcome. Life is not fair. It is tempting to believe that government can rectify what nature has spawned.
>
> (M. Friedman and R. Friedman, 1980, p. 168)

Thus he sees the attempt by government to create equality of outcome as a great mistake whether this has occurred in the post-war British Welfare State or in the Soviet Union. People's results or outcome should depend on their efforts and not on the initiative-sapping intervention of government. 'No arbitrary obstacles should prevent people from achieving those positions for which their talents fit them and which their values lead them to seek. Not birth, nationality, colour, religion, sex, nor any other irrelevant characteristic should determine the opportunities that are open to a person – only his abilities' (1980, p. 163). However, Friedman underestimates the amount of government action required to enable this equality of competitive opportunity to operate and his position therefore remains unclear.

Hayek believes, as does Friedman, that if 'equality' is pursued as an aim the net effect will be for certain group interests to dominate and prevail over others. If the state is making decisions that affect the distribution of wealth on the basis of a concept such as 'social justice' then it is only natural that people will band together into groups to try and influence the content of this infinitely variable 'social justice' and so get more of the rewards of state interference for themselves. 'In this tug-of-war between the various pressure groups which arises at this stage, it is by no means necessary that the interests of the poorest and most numerous groups should prevail' (Hayek, 1944, p. 86).

It is not Hayek's view that a redistribution towards the poor should be undertaken. This would be against economic growth and thus a brake on progress. He does not claim that the market provides just desert for everyone but that its overall effect is dynamic and expanding which will eventually filter through and create benefit all round. Brittan points out that Hayek denies 'that there is any moral merit attaching to the rewards people can obtain in the market place' (Brittan, 1980, p. 40). This is a message that is sometimes not very palatable to Conservative politicians who prefer to claim that the market will automatically reward people according to their value to society. Hayek accepts the inequality and lack of fairness in the market in the short term and demonstrates the economic role played by those fortunate enough to be rich. He provides four reasons why a rich stratum is needed in society. First to test out new products. The luxuries of today become commonplace tomorrow, the elimination of the less successful products will have occurred at no expense to the poor and the dynamic of the economy will have been fostered. Second, the rich can take risks setting up new enterprises, again ensuring progress and providing employment. Third, they can act as sponsors to the arts, education and research and finally they can act as an independent, powerful, check on oppressive government. Hayek does not believe that society should aim for equality of opportunity and he sees nothing wrong with, for example, inherited wealth which is needed for the efficient accumulation of capital. Equality of opportunity is impossible to achieve and raises all the same problems of definition and lobbying which he attaches to the concept of 'social justice'.

Finally, in this section on misconceived values, Hayek's views on democracy will be briefly mentioned. He focuses on this topic in the last volume of *Law, Legislation and Liberty*. His view is that majority decisions cannot be automatically seen as the best decisions. They may be worse than ones taken by individuals with greater knowledge of the facts and more concern for the effects of the decision. Thus democracy cannot be regarded as an absolute ideal but judged on the benefits it brings and at times it should be disregarded for some greater end. Hayek sees democracy failing as governments become more and more under the influence of powerful sectional interest groups. The more power government and an elected assembly has to distribute resources in society the more prone they are to pressure from such groups. It is in order to avoid this problem that Hayek proposes his model constitution. This constitution aims at ensuring the dominance of the 'rule of law', that is the general principles around which society is organised and to which government itself has to conform, and which also prevent the adverse influence of sectional pressure groups. There would be an upper assembly, 'the legislative assembly' for setting out the 'rules of law'. This would be made up of wise and mature people aged 45–60 who would serve for fifteen years and not be eligible for reselection. Each citizen would have one vote in their lifetime when they were aged 45. The whole aim of this procedure would be to ensure an assembly free from political or sectional pressure. There would then be a second assembly which would have to work within the constraints set out by the upper chamber. It would carry out the administration of government services, and be elected on principles much as those for the present House of Commons. It is not the details of such a constitution that matter but the implied attack on the contemporary processes of democracy.

Hayek believes that in order to preserve the stability and dynamic of society, the principles of democracy have to be relaxed because they lead to a battle between different interest groups who cannot all be satisfied and whose interests are mutually incompatible. Thus the conclusion of Hayek's recent argument is that certain basic issues have to be taken out of the democratic process, out of the bargaining arena of politics and government, and placed in the care of some body of responsible elders. As Gamble (1979a) has pointed out this notion clashes directly with the idea that democracy is the process of popular sovereignty, 'because it implies that there are many laws which should be beyond the power of government to alter, whereas the doctrine of popular sovereignty suggests that a government elected by the people has the right to overturn and refashion all laws' (1979a, p. 7).

Is the state necessary at all?

Having seen how Friedman and Hayek attack the notion of government intervention and the concepts that are used to justify it, this section will explore whether they think the state should perform any role at all. Friedman summarises his view of government in the following way:

> A government which maintained law and order, defined property rights, served as a means whereby we could modify property rights and other rules of the economic game, adjudicated disputes about the interpretation of the rules, enforced contracts, promoted competition, provided a monetary framework engaged in activities to counter technical monopolies and to overcome neighbourhood effects widely regarded as sufficiently important to justify government intervention, and which supplemented private charity and the private family in protecting the irresponsible, whether mad-

man or child – such a government would clearly have important functions to perform'

(M. Friedman, 1962, p. 34).

The first of the limited roles that Friedman accepts is that of 'rule-maker and umpire'. He believes that this is required because one person's freedom can conflict with another. He quotes a judge as putting this issue in the following terms: 'My freedom to move my fist must be limited by the proximity of your chin' (1962, p. 26). Friedman prefers conflicts over freedom to be resolved by voluntary arrangement (for example by private arbitrators) but accepts that this may not always be possible. However, over and above the role of government as arbitrator of unresolvable disputes, he agrees with Hayek that there should be a general agreement on the rules of the game so that the disputes don't arise in the first place. It is the role of government to ensure that these general rules are understood and complied with (1962, p. 25). The basic means by which this general agreement is achieved is through tradition and custom but Friedman sees the need for government to supplement this.

The mediating function of government is developed by Friedman in a discussion of property rights, which is of particular interest given the interventionist position of planning in relation to these rights. Friedman points out that there is no absolute definition of property rights, the exact extent of these rights being socially created. He uses the example of someone owning a piece of land and thus having the right to use this land as he thinks fit. But does this right extend to denying someone else the right to fly over it in an aeroplane? If so at what height and with what amount of noise? Thus one of the roles of government is to set out the general rules in relation to such issues by providing the exact definition of property rights. Once this has been done a minimal amount of interference with these defined rights is needed. Of course this does not exclude debate about the extent of these agreed rights in the first place, for example, debate on the scope and degree of development control required in society.

Closely related to the role of government in defining property rights is another one accepted by Friedman concerned with 'neighbourhood effects'. He describes how all transactions cannot be left to the market 'because of "external" or "neighbourhood" effects for which it is not feasible (i.e. it would cost too much) to compensate or charge the people affected; third parties have had involuntary exchanges imposed upon them' (1980, p. 51). As an example he talks about the effect of smoke: 'your furnace pours forth sooty smoke that dirties a third party's shirt collar'. The owner may be willing to pay the costs this imposes on the third parties but it is not feasible to identify all those affected. On another occasion he uses the example of the pollution of a stream (1962, p. 30). According to Friedman these cases of third party effects justify some degree of government intervention to remedy the adverse results. However, he points out that nearly all actions have some implications for third parties but this should not be used as an argument for greater and greater intervention. He is therefore at great pains to point out the limitations that should be placed on this principle. First, he describes how government intervention has its own problems. Beyond the basic point that all intervention will reduce individual freedom and so has a negative aspect, there is the problem that the costs and benefits of neighbourhood effects are incredibly difficult to assess, having been removed from the directness of market exchange. On top of this, government action will itself contain its own neighbourhood effects which may be unexpected or unrecorded. As a result government can often make matters worse rather than better. He gives an example of this in the US government housing programme which provides housing for the poor but which also has the detrimental effect of simply shifting the problem around and reducing the housing available. A British example might be the planning controls on office development in London in the 1960s which inflated the market value of existing premises and diverted investment into speculative office development (see for example, Taylor-Gooby in Taylor-Gooby and Dale, 1981, p. 59). Thus Friedman argues for an absolute minimum of government intervention to alleviate neighbourhood effects with the burden of proof over the necessity for intervention lying with government. If the effects of the market are not all that serious then, because of the additional neighbourhood effects created by government, it is better that the initial effects be accepted. On the whole he thinks that American society has gone too far with this kind of intervention and that many aspects would be better left to the market. He illustrates this with examples of roads and parks.

He accepts that for general access roads, with their many points of entry and exit, it would be impossible to collect money from all beneficiaries and therefore some form of government action such as petrol tax is unavoidable. However, he sees no reason for government to extend this to motorways where a direct market transaction is possible as it would be administratively feasible to identify users and set up toll booths to collect payment. Similarly a city park is used by many different people for many different reasons, including people who just look at it from their windows overlooking the park. It would therefore be impossible to collect payment according to benefit. However, for National Parks, which are at the moment regarded as a legitimate government responsibility, there is no such problem of identification and collection and therefore no reason why they could not be privatised.

Friedman also identifies another area which cannot be left to the market. This occurs when what he calls technical conditions make it necessary for a monopoly to operate. He does not define these technical conditions very precisely and in his more recent book (1980) seems to avoid discussing this aspect of intervention. He gives as an example of technical monopoly 'the telephone service' but spends more time in explaining why road, rail and postal services should not be considered to be in this category. Having tried to identify a technical monopoly he then suggests that there are three administrative possibilities, private monopoly, public monopoly and regulation. He is again vague about the occasions when some form of state involvement is necessary, through either regulation or the operation of an activity.

His last category of acceptable intervention is on 'paternalistic' grounds. In society certain kinds of people cannot be left to operate in the market because they are not 'responsible' and it is the duty of government therefore to protect such people. Having set out this principle he then confines its application to madmen and children. The latter he sees as being the responsibility of the family although in the last instance the family does not have open-ended freedom of action if this is against the interest of the child. Friedman accepts that having accepted the principle of 'paternalistic' government intervention there are no clear boundaries against extending the coverage. He believes that detrimental extension has occurred under the Welfare State. However, his means of defining the boundary of what is acceptable is also vague; he says 'there is no formula that can tell us where to stop.... We must put our faith, here as elsewhere, in a consensus reached by imperfect and biased men through free discussion and trial and error' (1962, p. 34). It is clear that he wishes to reduce paternalistic intervention to an absolute minimum and would, for example, not agree with the degree of paternalism advocated by Tories such as Gilmour.

How then does Friedman treat the question of poverty? According to him it should be tackled through private charity. He looks back with favour to the philanthropic activity of the nineteenth and early twentieth century, such as the hospitals, orphanages and 'settlement houses established throughout the nation to spread culture and education among the poor and to assist them in their daily problems' (1980, p. 172). However, he does accept that this charity will be insufficient and therefore government action is required as well. He thinks that a basic minimum for everyone in the community should be set and that government should ensure that no one falls below it.

Friedman's favoured means is through a negative income tax. However, he acknowledges that there is not going to be a clear definition of this minimum standard thus opening the door to too much intervention.

How does Hayek treat this question of the acceptable role of government? In cases where the market does need control he believes that 'the rule of law' should be used rather than government intervention. The 'rule of law' would set out the general principles of society in the same way as a game has a set of rules. The rules would evolve slowly through time and be generally accepted. It would mean that everyone knew exactly where they stood but, as in a game, the rules would not favour any particular person or group and no one would know beforehand who the winners were going to be. It is necessary that these rules are fixed and known beforehand and that government also conforms to them. Then 'within the known rules of the game the individual is free to pursue his personal ends and desires, certain that the powers of government will not be used deliberately to frustrate his efforts' (1944, p. 54).

Having established a social system which is dominated by economic market transactions and a general body of principles, 'the rule of law', which is over and above government, what exactly is there left for government do? The government has to ensure that the rule of law is maintained. In addition to such policing there are also the needs of national defence and action in the case of natural disasters. All these activities require government to obtain the necessary funding. These activities are carried out only because it is not efficient for the market to do so and the same applies to certain other services. These services include provision of information, such as land registers or statistics, and certain roads and civic amenities where it would be difficult to charge those who use them.

He also accepts that certain regulatory functions should be performed by government provided that there is no alternative, that the regulations are set out in advance, and that they do not rely on vague discretion. However, the boundary between acceptable and unacceptable regulation is not a clear one. He says that safety at work regulation is unacceptable except where the benefits can be shown to be large enough. Pure food laws, licensing of doctors and lawyers, and safety of public buildings would all be acceptable though there is no reason why bodies other than the government could not carry out the regulatory function. Another example he gives is that of building regulation which on the whole could be in the interest of public safety but 'wherever such regulations go beyond the requirement of minimum standards, and particularly where they tend to make what at the given time and place is the standard method the only permitted method, they can become serious obstructions to desirable economic developments' (Hayek, 1960, p. 355). Thus, on the regulatory function of government the general message seems to be that there are certain areas where this is permissible but within limits and only if there is no other means of carrying out the function. However, the criteria for setting limits on this form of action are not very clear.

In *The Constitution of Liberty* (1960) Hayek devotes a chapter to Housing and Town Planning. In this chapter he accepts that the conditions of urban life are complex and that this can lead to some need for a co-ordinating body and that there are instances when the property rights of individuals should be overridden. He accepts that this may lead to government action although preferring alternative agencies such as 'estate management'. Thus he accepts that in these urban situations the market is unlikely to cope with all the implications of individual actions and that a person's property will be affected by the actions of neighbours and any communal provision:

> In many respects the close contiguity of city life invalidates the assumptions underlying any simple division of property rights. In such conditions it is true only to a limited extent that whatever an owner does with his property will affect only him and nobody else. What economists call the 'neighbourhood effects' ... assume major importance. The usefulness of almost any piece of property in a city will in fact depend on what one's immediate neighbours do and in part on the communal services without which effective use of the land by separate owners would be nearly impossible.
>
> (Hayek, 1960, p. 341)

This can be compared to Friedman's discussion of 'neighbourhood effects'. On other occasions Hayek refers to the harmful effects of deforestation, some methods of farming and the smoke and noise from factories. However, he is at great pains to stress that any government action must be oriented towards ensuring that the market itself works better, rather than any attempt to supplant it. For him 'the issue is therefore not whether one ought or ought not to be for town planning but whether the measures to be used are to supplement and assist the market or to suspend it and put central direction in its place' (1960, p. 350).

Finally does Hayek believe that the state has a role in providing welfare? He says that there will always be inequalities in society that will be seen as unjust by those who suffer them. He goes on to claim that these inequalities are more easily borne if they are the result of the impersonal abstract forces of the market rather than consciously made decisions of authority (1944, p. 79). Having made this general point against the state taking on welfare functions to alleviate inequality, he accepts that there may be exceptions such as certain groups that cannot compete in the open market. His categories seem to be rather more extensive than those contained under Friedman's 'paternalistic grounds' and include, the old, the sick, handicapped, widows and orphans. He advocates the principle of a minimum income for such groups which should be granted only when the individual insurance principle has failed.

Bureaucratic interference

Both Friedman and Hayek mount a concerted attack on bureaucrats. This is of particular interest in the study of planning given that most planners operate as 'state bureaucrats' and also because of the amount of discretion that is currently inherent in the planning system. Friedman has commented on the post-war growth of a bureaucracy that has accumulated God-like powers and reduced self-help, self-respect and initiative. He sees this bureaucracy performing its duties on the basis of the false concepts mentioned above such as 'equality of outcome' and 'fairness'.

Hayek has also noted this growth and argued that there now exists in Britain an army of government officers who have never been elected and yet are able to impose their whims on the general public through their unregulated application of discretion. Hayek sees this arbitrary power as one of the main threats to society:

> 'it would scarcely be an exaggeration to say that the greatest danger to liberty today comes from the men who are most needed and most powerful in modern government, namely, the efficient expert administrators exclusively concerned with what they regard as the public good.
>
> (Hayek, 1960, p. 262)

If government itself were bound by the 'rule of law' this danger would be alleviated. There would be areas of life beyond the interference of government and bureaucrats would be challengeable.

The New Right has drawn upon the 'public choice school' which focuses on the role of bureaucrats (for a fuller account of this school see for example, Bosanquet, 1983, Chs 4 & 5; King, 1987, Ch. 6; Heald, 1983, Ch. 5; Green, 1987, Ch. 4). The views of this school have had a strong influence on the Thatcher governments. Schumpeter, writing in 1942, was one of the first people to raise the issue of the adverse relationship between bureaucrats and the efficient operation of capitalism. Most of the literature since then has drawn upon his notion of intellectuals using their control of ideas to influence and infuse the media and bureaucracies and as a result create a broad atmosphere that is hostile to the capitalist order. 'The social atmosphere ... explains why public policy grows more and more hostile to capitalist interests, eventually so much as to refuse on principle to take account of the requirements of the capitalist engine and to become a serious impediment to its functioning ' (Schumpeter, 1974, p. 158). Recently, these ideas have been taken up by the neo-Conservatives in the USA. For example, Kristol (1978) has talked about a 'new class' made up of public administrators in alliance with liberal intellectuals who have a 'hidden agenda' for the destruction of capitalism.

Friedman (1980) adopts this 'new class' thesis of Kristol. He attacks people who undermine the market through their support of the doctrine of equality and in a similar way to Schumpeter he points to 'government bureaucrats, academics whose research is supported by government funds or who are employed in government financed "think tanks", staffs of many so-called "general interest" or "public policy" groups, journalists and others in the communication industry' (1980, p. 174). He questions their claim to know and represent the 'public interest' and points to their high incomes as a demonstration of how well they have looked after their own interests. Even if it isn't their own interest that is promoted, the search for the 'public interest' will, in Friedman's eyes, always result in the fostering of other particular interests. Through claims to represent the public interest bureaucrats interpose themselves between the public and their political representatives and their actions only help certain groups. In a comment reminiscent of recent criticisms of planners, Friedman argues, 'higher level bureaucrats are past masters at the art of using red tape to delay and defeat proposals' (1980, p. 344) and they 'become persuaded that they are indispensable, that they know more about what should be done than uninformed voters or self-interested businessmen' (1980, p. 345).

So far the discussion of bureaucrats has concentrated on the effect of their values on undermining capitalism. However, there is another dimension to the public choice school's position which is concerned with the self-interest of bureaucrats and its effect on the economy. This aspect has been developed by the 'Virginia school' in the USA and the Institute of Economic Affairs in Britain (for details see for example, Bosanquet, 1983). The concern of these writers is that the imperatives of bureaucracy lead to economic inefficiency. The argument is that bureaucrats wish to foster their own utility and the growth of their empires. Higher salaries, reputation, quiet life, interesting work, control over the job – all these sentiments can be better achieved through expansion of the 'bureau'. This can become a greater and greater burden that has to be carried by the economy and one that is difficult to control, especially as the 'bureaux' are often in a monopoly position. On the basis that bureaucrats act in their own interest and need to be controlled, it has been suggested that there is a need to promote more competition within individual bureaucracies, to contract out functions to private firms, to reduce or contain salaries until voluntary redundancies create the required reduction in size, and even to withdraw the vote from bureaucrats and their families (Tullock, 1979).

Such government actions as privatisation and public sector redundancies can be seen as compatible with such views. It is also evident that the Thatcher government has an antagonism towards bureaucrats, seeing them as having a will of their own and opposing the government's own philosophy. Nigel Lawson has said, 'civil servants and middle class welfare administrators are far from the selfless platonic guardians of paternalistic mythology: they are a major and powerful interest group in their own right' (1980, p. 7). Meanwhile, Mrs Thatcher has expressed the antagonism rather more graphically:

> To believe that socialism is in some way morally superior to a free enterprise system is to believe that it is better for an official to take a decision than for an individual to take it for himself. What is more, one has to believe that bureaucrats, and the Socialist functionaries that direct them, are free from any normal human faults and actuated purely by a selfless desire for the public good.
>
> (Thatcher, 1977, p. 77)

So in implying that bureaucrats not only have motives of their own but also are directly associated with socialism and undermining capitalism, Mrs Thatcher is building on the views expressed by Schumpeter in the 1940s.

The political statement

Throughout the post-war period, apart from a brief moment of glory under Heath's 'competition policy', right-wing views remained on the side lines of politics until the late 1970s. By then the 'right' faction had gained sufficient power to dominate party policy. (An account of earlier political statements of the New Right philosophy can be found in, for example, Gamble, 1974; Kavanagh, 1987; or Norton and Aughey, 1981.) The rise in right-wing domination in the late 1970s built on the campaigning work carried out earlier in the decade. When the Conservative Party won the election in 1970 those who believed in the New Right ideas were keen to ensure that the manifesto ideas of 'competition policy' were not abandoned by the government once in office. In 1970 the Constitutional Book Club was set up to propagate the ideas of the New Right and 'reassess the values and aims of our society' and thereby 'expose the danger of "progressive" thinking which since the war has weakened both national enterprise and the unity and freedom of our society' (Boyson, 1970, p. 164). The first two volumes of this Book Club, both edited by Rhodes Boyson, were particularly influential (Boyson, 1970, 1971). The general message propagated by all the contributors to these volumes is that the activities of the state need to be severely reduced and replaced by an emphasis on individual initiative. The spirit of this antagonism between the individual and the state is graphically described by Boyson in the following way:

> A state which does for its citizens what they can do for themselves is an evil state; and a state that removes all choice and responsibility from its people and makes them like broiler hens will create the irresponsible society. In such an irresponsible society no one cares, no one saves, no one bothers – why should they when the state spends all its energies taking money from the energetic, successful and thrifty to give to the idle, the failures and the feckless?.
>
> (Boyson, 1971, p. 5)

The Constitutional Book Club writers claim that there is a need to re-establish the values of individual initiative that dominated Victorian Britain.

The assault on post-war values

As a result of Heath's U-turns the exponents of the New Right lost ground in the early 1970s, but were to reassert themselves when Mrs Thatcher won the leadership battle. This time particular emphasis was given to changing public opinion. The failure to hold ground after the 1970 election was seen as a failure to mount a sufficiently broad campaign to assert the values of the New Right. A major ideological battle had to be waged against the ingrained values of collectivism and socialism. Sir Keith Joseph was the leading light in this new campaign and he spent a lot of time in the late 1970s touring the university circuit putting forward his ideas. As Mrs Thatcher said in 1976,

> Our Conservative societies and clubs in the universities now have more members than those of other political parties. This is a development of the utmost importance. Those of us who believe passionately in a free society know that we must fire the imagination, and enthusiasm of young minds *today* if we are to safeguard freedom *tomorrow*.
>
> (Thatcher, 1977, p. 64).

The target audience was not just students. Sir Keith Joseph talked about the need to tap the common ground; to make the Conservative approach compatible with the everyday ideals and aspirations of the mass of the population. Prime importance was given at the time to ensuring that public opinion was supportive of the policies of the Right and Keith Joseph set up the Centre for Policy Studies to propagate these ideas as widely as possible. Mrs Thatcher echoed the need for a political strategy that not only pursued anti-state policies but also mounted a campaign against underlying values:

> a halt to further State control will not, on its own, restore our belief in ourselves, because something else is happening to this country. We are witnessing a deliberate attack on our values, a deliberate attack on those that wish to promote merit and excellence, a deliberate attack on our heritage and our great past.
>
> (Thatcher, 1977, p. 32)

So it is not surprising to find considerable attention given in the late 1970s to the inadequacy of the values that have supported state intervention in the post-war period and the superiority of the values propagated by the New Right. On the basis of this shift in values the more restricted role of the state is built.

The campaign to reassess the role of the state started with a critique of the increase in intervention after the war. According to Joseph the war 'not only further increased the actual role of the state, but also increased belief in the efficacy, indeed the virtual omnicompetence of state intervention' (1976a, p. 8). He puts this down to a number of reasons. First, the additional powers given to government during the war in the name of patriotism tend to perpetuate themselves as they are supported by vested interests. 'Second, war evokes the promise of a better future for all classes once war ends, a land fit for heroes, while simultaneously weakening the economy and burdening it with huge debts' (1976b, p. 51). He also points to the way in which the war created a feeling of solidarity which was conducive to collective beliefs such as those put forward by the Labour Party in 1945; 'the war's atmosphere of fellowship has frequently accompanied the expression of egalitarian beliefs' (Joseph and Sumption, 1979, p. 13) and so 'it was only natural that the socialist-Keynesian thesis on the capacity of government to solve social and economic problems should find the climate congenial' (Joseph, 1976a, p. 8). However, Joseph believes that this was misguided at the time and the continuation of this approach even more misguided. He gives two reasons for this continuation of state intervention. First its adoption by the Conservative Party itself and second its propagation by the opinion-formers in society. According to Joseph both these tendencies needed to be countered in the 1970s.

In a manner reminiscent of Kristol's attack on the 'new class' both Joseph and Thatcher stress the detrimental role played by intellectuals, the media and government bureaucrats in promoting the misguided path of increased state intervention. Mrs Thatcher illustrates this by pointing out the way in which an artificial belief in injustice has arisen 'cultivated by some philosophers, politicians, propagandists in the media and by social commentators', and that once the myth has taken hold then the next stage is 'the proposition that it is the job of government to intervene to correct unsatisfactory situations' (1977, p. 9). Once government has got a good footing in society then those reliant on it for a living will perpetuate its activities at the expense of the private sector, as Joseph puts it; 'expanded social studies departments, new professors, new research, new government departments all tend to generate pressures for more central powers at the expense of the business sector' (1976b, p. 51).

Joseph is particularly vociferous in pointing out that the 'middle way', originally advocated by Macmillan and others in the 1930s, has resulted in shifting the centre of gravity of the Conservative Party detrimentally to the left and contributed to the 'ratchet' of ever increasing socialism. Therefore part of the new crusade of shifting opinion has to be directed within the Conservative Party itself to defeat this 'middle ground' approach. He says, 'in retrospect, it is clear that the middle ground was not a secure base, but a slippery slope to socialism and state control' (1976b, p. 20). Then utilising a rather more poetic form he describes how

> the middle ground turned out to be like the will-o'-the-wisp, the light that flickers over marshland by night beguiling the weary traveller; as he moves towards it, the currents of air he sets up by his movement send it dancing away from him, and he goes on following, stumbling deeper and deeper into the mire.
>
> (Joseph, 1976b, p. 25)

He claims that the compromise of the middle ground is an artificial construct between politicians and has no relationship to the desires of the public. The net result of Joseph's analysis is that there is a need to mount a concerted campaign to counteract the influence of the 'new class' and the attitudes of 'middle ground' Conservatives. This has to be done in a way that taps popular feeling, providing the broad basis of support that was lacking in 1970, and tackles the issue at the fundamental level of changing the values that underlie state intervention.

It is a feature of the writing and speeches of this period that an attempt was made to show that capitalism is *morally* better than state intervention. The feeling was that socialism in the post-war period had captured the moral advantage and that this needed to be refuted. For example, Joseph in attacking the principle of redistribution said, 'redistribution is unwise. But it is also morally indefensible, misconceived in theory and repellent in practice' (Joseph and Sumption, 1979, p. 19). Thus the aim is to demonstrate that the values that support intervention are bad and need to be replaced by others. These 'good' values are self-help focused on the institution of the family, charity and voluntary action, competition, choice, creation of wealth and entrepreneurial spirit, acceptance of natural inequality and the merits of elitism. The 'proper' role of the state can then be built around these 'good' values.

Equality is the principal value that is seen to underpin the post-war interventionist state and therefore to require concerted attention. Thus,

> the persistent expansion of the role of the state and the relentless pursuit of equality has caused and is causing damage to our economy in a variety of ways. It is not the sole cause of what some have termed the 'British Sickness', but it is a major one.
>
> (Thatcher, 1977, p. 9)

The need to counter the appeal of equality is demonstrated by Joseph's decision to write a book (with Sumption) which seeks to refute the concept. Here, as elsewhere, the reliance on Hayek is amply demonstrated as all the arguments Hayek used to attack 'social justice', outlined above, reappear in the same form in Joseph's attack on 'equality'. Joseph and Sumption (1979) identify three reasons that are used to justify the redistributive function of the state. These reasons are: the elimination of poverty, equality of opportunity and a sense of community. They claim that the degree of poverty in Britain tends to be exaggerated because relative standards are used instead of absolute ones. Therefore a lot of what is treated as poverty is merely an expression of the fact that there are differences in wealth in society. If one accepts, as they do, that there is nothing wrong with having these differences then a lot of what is described as poverty really boils down to envy by those that have less than others. As Thatcher also says, 'Envy is clearly at work in the case of the egalitarian who resents the gap between himself and those who are better off' (1977, p. 4). Joseph and Sumption cite the case of antagonism towards the property developers in the early 1970s and put this down primarily to resentment that the developers made so much money while their critics failed to increase their own wealth (Joseph and Sumption, 1979, p. 17).

However, Joseph and Sumption do accept that a certain amount of real poverty does exist.

> By any absolute standard there is very little poverty in Britain today. There are those who, like the old, the disabled, widows and some one-parent families, have special needs. There are other cases of poverty of a kind which no society can entirely eliminate because they result from, say, gross mismanagement, alcoholism or some unforeseen disaster.
>
> (Joseph and Sumption, 1979, p. 28)

So this poverty stems from special needs or personal inadequacy.

Joseph and Sumption then look at the argument that redistribution can overcome inequality of opportunity such as that caused by differences in children's backgrounds. They claim that such arguments are not founded on fact and that the evidence seems to show that poor background provides greater incentive. They point to examples of people, such as Morris and Rockefeller, who came from poor environments and made fortunes. The third merit suggested for redistribution is that it creates a sense of community or brotherhood. It is claimed that whereas competition generates antagonism between people, state intervention fosters co-operation. Following Friedman and Hayek, Joseph and Sumption refute this with the argument that state intervention simply transfers the competition to a different arena, that of a battle for power and influence over government decision making.

So having sought to demolish the moral claims for the pursuit of equality and the legitimacy of using the state to try and achieve this end, Joseph concludes that any such attempt can lead to tyranny along Hayek's *The Road to Serfdom*. He says that we should not 'ignore the attitudes and motives that nourish egalitarian politics – pursuit of power, envy of those who are different, passion for domination over a battery-hen society – all adding up to a hatred of diversity. Of such motives are tyrants made' (1976b, p. 79). The argument is not that a free society has no place for those that believe in egalitarianism, it is rather that such people should not try and impose their views, through the state, on the rest of society. As both Thatcher and Friedman have said (Thatcher, 1977, p. 5; M. and R. Friedman, 1980, p. 175), if people believe in such principles they can go off and join a commune. Is this the appropriate place for planners?!

So what, in more detail, are the 'good' values that need to be fostered? After Adam Smith, the author most quoted by politicians of the New Right is Samuel Smiles (for example Boyson 1971, p. 6; Joseph, 1976b, p. 75; Thatcher, 1977, p. 97). This is a reflection of the importance given to the value of self-help. This value places great importance on the desire and ability of the individual to look after him/herself and the fostering of a sense of pride in overcoming problems and satisfying needs without seeking the help of others. It involves support and encouragement to the institution of the family as the best vehicle in which this self-reliance can operate. It also calls on people to take a caring and concerned attitude to those in the community that may require help. It therefore puts great value on voluntary and charitable actions. These principles are neatly summarised in Mrs Thatcher's definition of what she calls the 'healthy society':

> First, it is a society in which the vast majority of men and women are encouraged, and helped, to accept responsibility for themselves and their families, and to live their lives with the maximum of independence and self-reliance. Second, it is a society where everyone feels himself a responsible member of the community in which he lives and works; where he is inspired to play his part in ensuring the well-being of that community; and, in particular, where he shows a practical concern for those members who – for reasons of age, handicap or other disability – cannot fend for themselves without help. And, third, the healthy society is founded on the family. Family life is the bedrock on which the healthy society must be built.
>
> (Thatcher, 1977, p. 81)

Thus, the healthy society is one in which individuals, through increasing their concern for their own needs and also through greater charity to others, undertake an increasing responsibility for welfare functions in society and the state a reduced responsibility. As noted above, this reliance on the individual is seen to be superior not only in economic terms, that is in the Adam Smith sense that the sum of the play of individual interests leads to the best results overall, but also in moral terms. This morality is stated in terms of the freedom of the individual to choose and is counterposed to the socialist state philosophy which subjugates and directs people. It claims to place value on the dignity of the individual and express confidence in the ability of human nature.

The state in the post-war period through its massive intervention has taken away from the individual so many decisions and choices that morality has suffered. Mrs Thatcher has said:

> A moral being is one who exercises his own judgement in choice on matters great and small, bearing in mind their moral dimension, i.e. right and wrong. In so far as his right and duty to choose is taken away by the State, the party or the union, his moral faculties – his capacity for choice – atrophy, and he becomes a moral cripple.
>
> (quoted in Russel, 1978, p. 104)

One of the effects of this increased state activity, for example, in looking after those in need, is that people lose the desire to be charitable, to make their own personal sacrifices and decisions to help their fellow men and women.

One of the attitudes that the New Right seeks to change is that which sees this self-help principle, operating in the context of capitalism, leading to selfishness. Also, the attitude that there is something immoral in the desire to make money (see for example, Thatcher, 1977, p. 88). They seek to show that there is nothing wrong in the pursuit of wealth and profit. The entrepreneurial spirit is to be praised and is the cornerstone of a thriving economy. Competition provides the incentive to work hard and ultimately the buoyant economy benefits everyone in society through the progress that is engendered.

This same spirit is generated through the acceptance of inequality. It is considered that efforts to counter the natural human condition of inequality will lead only to the removal of the incentives to use one's talents and attributes to the full and as a result society as a whole will suffer. This argument is used by Joseph, as it was by Hayek, to justify inherited wealth. It is also used to support the profit motive and Joseph specifically mentions that profits from property are beneficial.

Following Hayek again, Joseph (1976b) points to the complexity of society as another reason for preferring decision to be taken by individuals and families rather than by bureaucrats in the remoteness of their offices.

> No civil servant, however clever, can respond centrally nearly as sensitively or effectively to the endlessly shifting changes of home and world demand and supply as individuals do, whose careers, families and savings depend upon their being right more often than wrong in judging the market.
>
> (Joseph, 1976b, p. 76)

Thus, whatever its faults, Joseph sees the market as far preferable to bureaucratic decision-making which is faced with the impossible task of making decisions in a highly complex situation and lacking any reasonable criteria of choice. As he puts it:

> the blind, unplanned, unco-ordinated wisdom of the market is overwhelmingly superior to the well-researched, rational, systematic, well-meaning, co-operative, science-based, forward-looking, statistically respectable plans of governments, bureaucracies and international organizations preserved from human error and made thoroughly respectable by the employment of numerous computers.
>
> (Joseph, 1976b, p. 57)

Re-orienting the state

So what does this shift in values and the prominence given to market processes imply for the role of the state? Gamble has described one of the purposes of Thatcherism as the 'reorganizing of the state sector in a period of recession' (1979a, p. 4). He illustrates how this reorganization involves the rolling back of the state in some areas and the strengthening of its role in other areas. This differential approach to aspects of state activity is epitomised by reductions in welfare provision and increases in law and order. Is this reflected in the statements of Thatcher and Joseph? They are always at great pains to point out that they do not believe in *laissez-faire* and that there is a role for the government. Joseph (1978) has described this role in the following way: 'governments can help hold the ring, provide infrastructure, maintain a stable currency, a framework of laws, implementation of law and order, provision of a safety net, defence of property rights and all other rights involved in the economic process' (1978, p. 20). Meanwhile, Mrs Thatcher sees government 'as performing those functions – defence, maintaining law and order, administrative justice, preserving a stable currency, and providing collective services (sewage, traffic control) which private enterprise cannot conveniently supply' (quoted in Russel, 1978, p. 15).

These statements do suggest that there are certain activities of government that need to be retained and indeed strengthened and other areas where a cutting back on state activity is required, either completely or in some cases by allowing the private sector to operate alongside government. The areas of 'rolling back' seem to fall into two categories, that of economic activity and that of welfare activity.

There is no doubt that the state is seen to have an important role in maintaining law and order and defending the country. 'The first duty of any government is to safeguard its people against external aggression; to guarantee the survival of our way of life' (Thatcher, 1977, p. 41). This duty relates not only to law and order but also to the whole legal system within which the market operates, including the laws of property. The government is also seen to have a role in maintaining a stable currency. These functions can be summarised as providing the social stability, strong legal framework and financial security within which the individual actions of the market process can confidently take place. The more secure government can make this framework the easier it will be for entrepreneurs to take risks and the more likely that the economy will recover.

However, as mentioned in the last chapter, Joseph believes that this confidence for risk-taking will occur only if the state sector is drastically reduced. There is a need to reduce state expenditure and taxation in order to provide the necessary boost to the private sector. He complains that since the war we have witnessed the biggest government spending spree of all time.

So part of government's role in providing the right context for the market is to reduce state expenditure and artificial involvement in industry through subsidy or state enterprise. There is another important requirement. This is the reduction of controls and interference on enterprise: 'the private sector that produces the goods which people want is restricted by controls, over-taxed by local and central government and harassed by officials' (Joseph, 1976a, pp. 18–19). This need to reduce controls and regulations is often linked to small businesses (for example Thatcher, 1977, p. 34). The argument is that they are particularly affected by all the red tape and restriction of the post-war 'corporatism' in which the needs of the large firms have dominated. There is therefore a particular need to lift these restrictions and allow small firms to flourish again as the seedbed of a growing economy. However, as mentioned earlier, there is also a moral argument to this freedom from government control and regulation.

From the statements of Thatcher and Joseph it would appear that another area where positive action by the government is acceptable is in providing infrastructure. This area is obviously of particular relevance to town planning but unfortunately they do not expand on their statements. It is therefore difficult to judge what might be included under infrastructure. Is it simply sewers and traffic control, or does it encompass a much broader definition, say to include a range of shopping and housing? They do not even go as far as Friedman in distinguishing between different kinds of infrastructure such as city or national parks, local or national roads. Mrs Thatcher was specific on one occasion when she said that she would not include the protection of essential shops in an area as necessary state-provided infrastructure. She criticises the attempt by the West Midlands Council to gain the powers to provide state-owned shops where certain services were lacking. The only principle that one can deduce from their speeches is that if infrastructure can be provided by the private sector, then the state has no role. However, this does not help with the question of the quality of that provision, as the private provision may be uncertain, badly located or oriented to particular clients.

The final aspect of government activity that needs to be discussed is that of welfare provision. It is clear that this is an area where it is thought that the state should be 'rolled back'. This needs to be done not only for financial and economic reasons but also because to cut back on state provision reasserts the moral values of self-reliance and charity mentioned earlier. There are some areas that can be removed from state control altogether and others where private provision needs to be generated as an alternative alongside the state. There is an acceptance that in society some people cannot pay for their own needs, for example through insurance, and that the state has to take some responsibility. Joseph has said that 'some special groups, such as widows and the disabled, have too little income in general for their special needs' (1976b, p. 61). On another occasion he has included elderly people and some one-person households (Joseph and Sumption, 1979). Mrs Thatcher has also referred to those with genuine needs such as elderly and disabled people and one-parent families (1977, p. 84). Although it is accepted that there are those who cannot provide for themselves in the market, the answer to the problem is not necessarily seen as direct provision of services by the state. Rather the solution is sought in providing such people with the means whereby they can purchase their requirements in the market-place through some form of negative income tax or tax credit system.

At her first Conservative Party Conference after winning the leadership contest, Mrs Thatcher gave delegates her vision of the future society that the Conservative Party should strive to achieve. It is worth recalling that vision to conclude this section:

> Let me give you my vision: a man's right to work as he will, to spend what he earns, to own property, to have the State as servant and not as master; these are the British inheritance. They are the essence of a free country, and on that freedom all our other freedoms depend.
>
> But we want a free economy, not only because it guarantees our liberties, but also because it is the best way of creating wealth and prosperity for the whole country, and it is this prosperity alone which can give us the resources for better services for the community, better services for those in need.
>
> By their attack on private enterprise, the Labour government has made certain that there will be next-to-nothing available for improvements in our social services over the next few years. We must get private enterprise back on the road to recovery, not merely to give people more of their own money to spend as they choose, but to have more money to help the old, the sick and the handicapped.
>
> (Thatcher, 1977, p. 33)

Conclusions

A major campaign to challenge the values of the post-war consensus was launched in the late 1970s and this marked an important change from the Heath to the Thatcher periods. A shift in popular attitudes was seen as a necessary basis for reorienting the role of the state it-self. This ideological campaign drew heavily on the work of academic writers especially Friedman and Hayek.

The reorientation of the state is based on the view that a market system of decision-making is always preferable to any political or bureaucratic process. Any residual functions of the state should sup-port or supplement the market. Attempts by the state to try and control the vast complexity of society are doomed to failure and the state is bound to make innaccurate and simplified assumptions that will create unforeseen effects and be subjected to unsatisfactory pressures. Only the market mechanism can cope with this complex-ity. The emphasis on market processes incorporates an attack on any principles that might be employed as an alternative basis for deci-sions. Thus both Friedman and Hayek devote considerable energy to demonstrating the meaninglessness of concepts such as 'equality' or 'social justice' and hence the impossibility of using these concepts as a justification for intervention in the market process. Their argu-ments are adopted by Keith Joseph who attacks the attempts of post-war governments to achieve greater equality in society. Having shown how previously-held values are misguided, the political strate-gy is to then replace them with an alternative set of values that are supportive of a market-oriented system of decision-making. These new values include self-help, initiative, family responsibility, and charity.

The reorientation of the state involves an attack on bureaucrats who are seen as central to the old Welfare State. They use just the kind of concepts, such as the 'public interest', that are attacked as meaningless. Politicians in the post-war period, of whatever party, are also seen to have adopted what Hayek calls the 'socialist' aims of the Welfare State based on 'social justice'. The market needs to be freed from these political and bureaucratic constraints so that econ-omic progress can be achieved.

However, Hayek and Friedman both accept that the market is not perfect and that the state does have some limited role to play. They agree that certain 'bad neighbour effects' cannot be controlled through the normal market process and that in such cases the state has a role. However, they also point out that state intervention can have unintended side effects and therefore it cannot be assumed that the state should always intervene when the market operates ineffi-ciently. Sometimes the inefficiency of the market is preferable to the even greater inefficiencies of intervention. It is also agreed that the state might have a 'paternalistic' role in providing for those that are unable for one reason or another to look after their own interests in the market-place.

The views of both Friedman and Hayek are presented with

strength and conviction but they do contain a considerable amount of vagueness. For example, in defining the exact extent of property rights, deciding when to intervene to ameliorate neighbourhood effects or for paternalistic reasons. These boundaries are open to debate and it is possible to determine Friedman or Hayek's exact position only when they use concrete examples. Any government trying to adopt their ideas has considerable flexibility of interpretation.

The general lines of argument presented by Friedman and Hayek are adopted by Joseph and Thatcher in their own reorientation of the state. They propose that some aspects of state activity should be retained and even strengthened while others should be reduced. The state should be strengthened in its support for the market through maintaining social stability and the appropriate legal and financial framework. They say that support should also include the provision of infrastructure although it is difficult to determine how far they would extend this particular role. They want to reduce the state's role in both economic activity and welfare activity. In economic activity the state should interfere less in the market, refrain from subsidy, and reduce the regulatory burden on firms. The government should also seek to reduce the tax burden on the economy and hence scale down its activities and cut public expenditure. A major aspect of this cut in expenditure would come through the reduction of state activity in welfare provision. The onus would shift to individuals through insurance, family responsibility and charity. Following Friedman and Hayek, the state would still be required to provide for certain limited categories of people in need.

Another idea elaborated by Hayek is the 'rule of law'. This is seen as deriving from tradition and custom and gaining a general consensus of agreement and standing over and above government. The 'rule of law' is supportive of a market system and property rights and has to treat all people in exactly the same way (see Hayek, 1960, Ch. 15). Thus any government that attempts to interfere detrimentally in the market system or seeks, through some notion of distributive justice, to counter the unequal effects of the market by action directed at particular people will be acting against the 'rule of law'. Hayek identifies one of the problems of post-war society as the confusion of the 'rule of law' with legislation 'made' by government. The danger here is that governments expand their power and are prone to pressure from sectional interest groups. This leads Hayek on to a critique of democracy.

However, there appears to be a contradiction in Hayek's approach. Alongside the praise of the free market he gives overriding importance to the 'rules of law' derived from custom and tradition. According to Brittan (1980), Hayek's views are based on two different political philosophies: classical liberalism, built on limited government and free markets; and conservatism which stresses tradition and the hidden wisdom of existing institutions. This dual nature of his philosophy means that a selective reading of his ideas can be used to support the different positions of liberalism or authoritarian conservatism. However, as described in the last chapter, one of the distinctive features of Thatcherism is just this combination of philosophies and the manner in which the inherent contradictions are resolved is one of its unique attributes. Thus there is much similarity between Hayek's philosophical position and that of Thatcherism.

Thus certain tensions are created by the coexistence of economic liberalism and authoritarianism. As Hayek puts it:

> It is part of the liberal attitude to assume that, especially in the economic field, the self-regulating forces of the market will somehow bring about the required adjustments to new conditions, although no one can foretell how they will do this in a particular instance.... The Conservative feels safe and content only if he is assured that some higher wisdom watches and supervises change, only if he knows that some authority is charged with keeping the change 'orderly'.
>
> (Hayek, 1960, p. 400).

This Conservative tendency can be observed in the authoritarian strand of Thatcherism. As will be seen in later chapters, the application of these principles has led to a severe reduction in the power of local government and an increase in the centralisation of government activity (Krieger, 1986). This is clearly evident in the trends in local government finance and the abolition of the Metropolitan Counties (McAuslan, 1981). This centralisation has implied a disregard for democracy at the local level which again reflects authoritarian attitudes. Increased power has accrued to central government and much of this power has been used to deregulate the market, undertake privatisation and other measures which increase the dominance of the market mechanism. Nevertheless the concentration of government power does create tension with the liberal preference for minimal government and generates a potential conflict with Hayek's 'rule of law' which contains the inherent aim of restricting the scope of government.

Though some mention is made of town planning by the academic writers covered in this chapter, the topic is not one that receives a great deal of attention. However, some of the above themes have a particular relevance. As indicated in Chapter Two, there is a strong tradition that identifies planning activity with safeguarding the public interest. Thus the antagonism to the use of concepts such as 'social justice' and 'fairness' could relate to planning. It raises the question of whether a justification of planning based on such concepts is 'meaningless'. Many town planners would identify with Hayek's category of 'expert administrators looking after the public good' which he believes has created so much damage to society. The attack on bureaucrats and their use of these concepts can clearly be related to planners working in central and local government. Can they be said to be part of the 'new class' and hence in opposition to the market? Another aspect of planners' role is their administrative discretion. In their regulatory function of development control do they administer discretion resulting in random variations, ambiguity and uncertainty? If so, can they be said to be in opposition to the 'rule of law'? Later chapters explore the way in which the government's actions since 1979 reflect the Thatcherite response to these questions. First a conceptual framework is briefly outlined to help establish the links between ideological themes and planning practice.

Conceptual framework

The various themes are drawn together under three headings. However, it is accepted that there is considerable overlap between the three headings and therefore the framework will be used flexibly.

Principles of decision-making

In Chapter Three two basic strands were identified that give the Thatcher ideology its distinctiveness. These strands were labelled 'neo-liberal' and 'authoritarian'. This ideology affects both the **principles** of decision-making and the **procedures** of decision-making. There has been a clear shift in the principles used in determining decisions towards those based on the neo-liberal strand of the ideology. As a result market criteria dominate and there is a reduction in the opportunities to employ alternative criteria based on social or community need. In support of this approach it is argued that society is highly complex requiring a decision-making process based on the market. Any other approach can lead only to distortion, political bias and economic sluggishness.

Procedures of decision-making

Turning to the procedures of decision-making one can observe the way the authoritarian strand of Thatcherism comes into play. Thatcherism destroys the corporatism that was the basis of the post-war consensus and imposes a style of government that we will call 'authoritarian decentralism'. This involves withdrawing power from local government and placing it in the hands of the central state. This centralised power is then used to establish a mechanism for ensuring that the details of decision-making take place in the decentralised market-place. Democracy and involvement, other than in the market or general elections, are downgraded in order to ease this process.

Anti-bureaucratic sentiments

These changes are supported by another aspect of Thatcherism which we will term 'anti-bureaucratic sentiments'. The attack on bureaucrats forms part of the attempt to establish the dominance of market criterea. Bureaucrats are blamed for justifying their intervention on the basis of meaningless concepts in order to maintain their position of power. This attack downgrades the use of all non-market criteria. Hayek's emphasis on the 'rule of law' aims at restraining the power of government and also seeks to reduce the degree of administrative discretion. An overall legal framework would remove what he has called the 'administrative despotism' of planners. These anti-bureaucratic sentiments have been used as part of the populist programme of Thatcherism.

Chapter five

New Right thinking; planning under siege

This chapter explores in some detail the right-wing views of planning that were emerging during the 1960s and 1970s to challenge the postwar consensus. These views, which were radical in the context of the period, gained ground from the late 1960s onward. The purpose of the chapter is to identify whether there is any relationship between these radical views and the Thatcherite position identified in the last two chapters. Having established the extent of this relationship at the level of ideas, later chapters will examine whether this body of thought can be detected in the changes to the planning system since 1979. The material for this chapter is chosen on the basis of its influence on the right-wing of the Conservative Party and its leadership. The chapter is divided into three sections; the first explores the early influences from the USA with particular emphasis on Jane Jacobs and the 'Houston experience', the second investigates the way in which these ideas have been incorporated into British thought with an emphasis on the writings of the Institute of Economic Affairs and the Department of Land Economy at Cambridge, and the third section covers the efforts, such as those of the Adam Smith Institute, to keep up the pressure since 1979. Finally the conclusions will draw out the common threads from all these contributions and relate them to the Thatcherist ideology utilising the conceptual framework outlined in the last chapter.

The influence of the USA

Jacobs and Banfield – two early voices of inspiration

Sir Keith Joseph's Centre for Policy Studies produced a *A Bibliography of Freedom* (1980) which sought to counteract the dominance of 'collectivist' thinking in contemporary debates and academic reading lists. In this bibliography there is a short section of sixteen references under the heading of 'Urban Policy'. No less than half of these are American books indicating the strong influence of ideas from across the Atlantic on New Right thinkers in the urban field. The two most influential American authors included in the bibliography will be reviewed focusing on their attitudes to the role of planning.

Two books by Jane Jacobs get a mention: *The Death and Life of Great American Cities* (1965) and *The Economy of Cities* (1970). The latter also appears on Sir Keith Joseph's reading list which he circulated to senior civil servants in the Department of Industry when he took over tenure of that department. The books are described in the bibliography as 'two brilliant studies of the failure of central planning and the virtues of the decentralised and spontaneous market order' (Centre for Policy Studies, p. 21). So although both books were written in the 1960s they are obviously considered to have an important and relevant message for planning today. It is worth spending some time examining Jacobs' views as these have been widely disseminated and have influenced many other writers.

The Death and Life of Great American Cities was first published in 1961 and is largely a reaction to the comprehensive redevelopment schemes and urban motorways that dominated urban policy at that time. However, the book also contains a broader message about planning and the opening sentence reads: 'This book is an attack on current city planning and rebuilding, (Jacobs, 1965, p. 13). According to the author planning has completely failed to stem the decay of some American cities and has even contributed to their economic decline. The problem is that planners have had no new ideas for a generation. Jacobs sees contemporary planners being guided by Howard and Corbusier and, although not adopting the Garden and Radiant City solutions, being still imbued with their underlying principles. These principles she categorises as self-containment, the inward-looking neighbourhood, decentralisation, grass instead of streets, control of commercial activity and the wish to sort out uses into separate land-use zones, for example cultural and public functions. She describes this last aim as one of seeking to isolate certain uses from the contaminating influences of the rest of the city.

The reason Jacobs gives for the adoption of these 'false' principles by planners is their inability to understand and cope with the complexity that makes up city life. Because of this inability they are forced into meaningless simplifications (1965, p. 43). In their effort to come to grips superhumanly with this complexity, planners have resorted to concepts that have no relation to real-life cities. Most of Jacobs' energies are devoted to demonstrating the nature of this real world from which planners hide. She contrasts the appeal of everyday life in the city with the sterile and monotonous world devised by planners.

She criticises planners for being guided by 'principles derived from the behaviour and appearance of, towns, suburbs, tuberculosis sanatoria, fairs, and imaginary dream cities – from anything but cities themselves' (1965, p. 16). She urges planners, rather than trying to pretend that they are omnipotent, to utilise the knowledge of residents themselves who do have detailed knowledge of their neighbourhoods.

The key to overcoming planners' stultifying concepts, according to Jacobs, is to stress diversity as the ingredient that makes cities exciting places to live in (1965, p. 24). Cities are the home of an enormous variety of small business and retail outlets, ranging from supermarkets to specialist grocers, and cultural provision. She claims that one of the major destroyers of diversity is the way planners analyse each land use in turn thus avoiding the essence of the city which is the interplay between uses. She suggests that the mixed complexity of uses must be faced head on. She discounts planners' apprehension over mixed-use zoning claiming that such fears are generally unfounded. First she examines whether such zoning will invite harmful uses such as scrap yards. This she discounts on the grounds that, if the area is diverse and thriving, market forces will ensure that harmful uses will go elsewhere where it is cheaper to locate. She then examines other uses that planners often object to, such as night clubs, offices or industry. These she claims are not harmful but necessary for creating diversity. Any problems such as pollution or noise are better dealt with through legislation directly geared to these problems rather than through land use regulation. She does however accept that there are some uses that are harmful. 'They can be numbered on one hand; parking lots, large or heavy trucking depots, gas stations, gigantic outdoor advertising, and enterprises which are harmful not because of their nature exactly, but because *in certain streets* their scale is wrong' (1965, p. 246). These uses can usually afford high rents and are therefore a potential problem in city areas and can have a destructive influence. She is particularly concerned about the effects of the last category, i.e. uses of inappropriate scale, and she suggests that what is needed is not zoning on the basis of use but controls on the permitted street frontage allowed.

However, even if planners were reformed and new regulations were devised, she accepts that there are other problems in maintaining diversity. One problem derives from the financing of development. She sees the major influences behind city change as the insurance companies, banks, pension funds and to a lesser extent government agencies. Such institutions are conservative and tend to limit their investment to safe areas and reject the poorer neighbourhoods. They also like to invest in large schemes which have a detrimental effect on the diversity of the city leading to monotonous and single-use blocks. She also talks about how successful diversity can bring about its own downfall. Once the diversity and excitement makes an area popular then certain uses will want to locate and expand there and will have greater economic means of doing so. After a while those uses which cannot afford the same rent levels will be pushed out and the area will lose its mixed use characteristic and become dull and monotonous again. She describes how banks, insurance companies and prestige offices are the worst offenders at generating this process.

So what does Jacobs suggest can be done to ensure diversity in cities? Her starting-point is to see individuals and their actions as being the source of all that is good in cities. Through their myriad separate actions comes the necessary diversity (1965, p. 255). The role of public action is to facilitate this private sphere.

> The main responsibility of city planning and design should be to develop – as far as public policy and action can do so – cities that are congenial places for this great range of unofficial plans, ideas, and opportunities to flourish, along with the flourishing of the public enterprises.
>
> (Jacobs, 1965, p. 255)

So what means should be adopted to encourage these private plans and what role is there for public planning?

She proposes quite a positive role for planning though not along the lines of post-war practice. She sees planners intervening in order to maintain diversity. The reason for this intervention could be to create diversity where it has not arisen or to ensure that existing diversity is maintained against the self-destructing forces mentioned above. She mentions a number of tactics that planners could adopt in promoting diversity. One is to zone positively for mixed uses and diversity. This may be done through insisting on the retention of a variety of building types and ages that will command different rent and therefore different uses. Public and quasi-public bodies should locate their uses in areas where they can add to the diversity and then stay there, whatever the rise in property values that results from the increased success of the area. Planners should also encourage as many areas of diversity as possible thus reducing the chance that a few might become overpopular with the users that can afford high rents. In slum areas people, either tenants or landlords, should be given subsidies so that they themselves can improve their housing conditions without having to destroy the area. The problems of cars should be dealt with through small piecemeal traffic management rather than urban motorways. Planners should give up their ideas of grand visual order and concentrate on small-scale schemes like tree planting and small individual landmarks (Jacobs, 1965, p. 387). Lastly planning should be administered through small units of about 50,000 to 200,000 population to ensure a full knowledge of conditions in the area and facilitate public involvement.

The purpose of Jacobs' later book *The Economy of Cities* is to explore the reasons why some cities grow while other stagnate or decline. She makes only a few passing references to planning. Her main message is again focused on the concept of diversity. This time the argument is that for healthy city development you need a diverse structure of enterprises and in particular small enterprises. She sees the core of the city growth process as the addition of new kinds of work to the old established enterprises. A city grows through the diversification and differentiation of its economy, occurring in a gradual piecemeal way. In so doing it is continually creating new exports to take the place of the old.

One of the broader messages for planners from her thesis is that this growth process can take place only if the city is inefficient (1970, pp. 96–7). City growth depends on a large number of small firms with plenty of opportunity for 'trial and error' as this allows for the easy birth of new enterprises. The availability of cheap accommodation and a wide range of goods and services are required. A physical arrangement that is efficient for established firms, which are usually large, will not create the conditions for diversity needed for the growing firms. In fact the very presence of these small firms gets in the way of the efficiency of the larger and more established ones. Not only do small firms cause physical problems but also they produce competition of goods and labour markets. If the planner tries to improve the efficiency of the city in such a way that benefits these established and larger firms she/he will be contributing to the decline of that city as the dynamic sector of the economy will be pushed out.

Apart from this general message Jacobs makes two particular references to planning. In one of these she criticises the way planners categorise different classes of industry. She claims that these categories have no correlation to the processes that contribute to city growth and that the application of these planner's categories can stifle development and reduce growth. A second direct reference to planning comes in her attack on policies that attempt to locate industry in rural areas. She draws attention to the failure of the Chinese attempt to graft industry on to villages and some American companies' attempts to locate new industry in rural parts of India. Her argument is that the relocated industries lack the necessary linkages and support system to prosper and contribute to growth. She applies the same argument to the British New Towns policy which she sees as a misguided initiative that ignored the dynamic growth-producing characteristics that the expanding centres of London and Birmingham had to offer.

Her only other reference to planning comes in her account of the history of Pittsburgh, used as an example of a declining city. The economy of this city began to stagnate around 1910 when it became reliant on a few large well-established companies. When these declined there was no potential in the city for replacement and people began to leave for work elsewhere. The composition of the population became heavily weighted towards the very young and old. Since the Second World War all kinds of interventionist policies were devised to try and redress the situation but with no success. This brought into the city:

> thousands of economic consultants, industrial analysts, regional planners, city planners, highway planners, parking planners, cultural planners, educational planners, planning co-ordinators, urban designers, housers, social engineers, civic organizers, sociologists, statisticians, political scientists, home economists, citizen-liaison experts, municipal-service experts, retail-trade experts, anti-pollution experts, publicity experts, development experts, redevelopment experts, dispensers of birth control pills to the poor, and of course experts in attracting industry. They have industriously documented, studied, analysed, psychoanalysed, measured, manipulated, cleaned, face-lifted, rebuilt, cajoled, exhorted and publicized Pittsburgh.
>
> (Jacobs, 1970, p. 199)

As a result expensive urban renewal and highway programmes have occurred but the city is now (1969) economically worse off than before. Jacobs concludes that 'artificial symptoms of prosperity or a "good image" do not revitalize a city'. What is needed, she claims, is the direct regeneration of the economic growth processes, that is through developing a variety of small enterprises, to create the export potential that can be extracted from the existing economic foundation, however depleted that may be.

So what conclusions can one draw about Jacobs' view of the role of planning? There is no doubt that she has very harsh words to direct at planners and the simplistic principles that they try to impose on society. However, a careful reading of her work does not seem fully to support the view that it can be simply regarded as a presentation of 'the failure of central planning and the virtues of the decentralised and spontaneous market order' as claimed by the Centre for Policy Studies. Her position seems to be more complex. She points out that the 'spontaneous market' or diversity as she calls it has its own problems. The desirable outcome can be destroyed by those that have most money in the land market, such as banks and insurance companies. Thus if diversity is to be maintained some kind of control over the market is required. So while there is an element of 'freeing up the market from planners' restrictions' in her work this is not a blanket 'the market knows best' approach. Rather she appears to be saying that within the market processes there are some aspects that are beneficial and which are being held back by restrictions, but that equally there are detrimental market forces and these may still require some form of public intervention to ensure the 'best' outcome. 'Best' is interpreted as a variety of land use and economic activity.

Another book included on the Centre for Policy Studies' Bibliography is Banfield's *Unheavenly City Revisited* (1974). This is a revision, published in 1974, of a book written four years earlier after the period of unrest and rioting in American cities. The book attracted a lot of media attention and influenced the Nixon administration and the author served on the Model Cities Programme. The 'neo-Conservative' Kristol described the work as 'easily the most enlightening book that has been written about the "urban crisis" in the U.S. It is part of a healthy revisionist trend in the social sciences' (1970, p. 197). However, this was not an opinion shared by most of the academic establishment. Nearly all the reviews and comments on the book were condemning. A typical description stated that 'it is often wrong in its assumptions, fallacious in its reasoning, warped in its values, wild in its conclusions, and dangerous in its prescriptions' (Lockard, 1971). Banfield's reaction to such comment was that the authors were politically motivated and did not read his book with any care and attention (for example Banfield, 1971a, 1971b). So what were the arguments in the book that generated so much controversy and interest?

In the introduction Banfield says 'my conjecture is that owing to the nature of man and society (more particularly, American culture and institutions) we cannot "solve" our serious problems by rational management. Indeed by trying we are almost certain to make matters worse' (1974, p.ix). In pursuing this argument he first defines what these 'serious' problems are. He distinguishes between the problems that 'threaten the essential welfare of society' and those which he calls issues of 'comfort, convenience, amenity and business advantage' (1974, p. 9). He claims it is the latter category that planners seem to spend all their time and society's money trying to solve. To him this is unnecessary because the problems are either unimportant or will take care of themselves; indeed interference can often make matters worse. For example he mentions that the ugliness of American cities is a pity but not a disaster and unlikely to affect the humanization of society. Also the effort of planners to prevent the decline of central business districts may be misguided because overall wealth will be increased by the move to the suburbs.

The major problems that Banfield *does* see as important are those that threaten to disrupt the stability of society or, as he puts it, those problems that affect the 'essential welfare of individuals or the good health of society' (1974, p. 9). He admits to finding it difficult to define these problems precisely but he includes not dying before one's time, not seriously impairing people's health, not unhappily wasting one's powers, maintaining society as a going concern and its free and democratic character, and the production of desirable human types. In essence these problems boil down to those of crime, poverty, ignorance and racial injustices. He sees the biggest threat to the health of society coming from enclaves of such problems in which people develop a 'separateness' and solidarity that hampers progress and is a threat to order. For Banfield the cause of such problems is the existence of what he calls a lower class who have often migrated to the cities from rural areas and who have a harmful culture which he calls 'present orientedness'. His view is that 'so long as the city contains a sizable lower class, nothing basic can be done about its most serious problems' (1974, p. 234). No amount of money put into schools, housing or job promotion will have any effect. Instead the lower class will have to disappear through acquiring the attitudes, motivations and habits of the rest of society. The general message, then, is that government intervention wastes money and time on schemes that are geared to insignificant problems that will take care of themselves anyway or on people who can never be helped until their attitudes change.

Another dimension of Banfield's argument is that the middle class and government officials have accentuated the problem of the city. Through their cultural ideal of 'service' and 'responsibility to the community' they have led public opinion into thinking that government programmes can solve the problems. This has given the lower class false hopes and also resulted in an attitude of blaming society rather than individual attitudes. As the critics (J. Friedman, 1971, p. 122; Lockard, 1971; Rossi, 1971) have pointed out Banfield discounts the existence of, for example, institutional racism or social forces that create the separate culture. Instead he attacks views that see 'poverty as lack of income and material resources (something external to the individual) rather than inability or unwillingness to take account of the future or to control impulses (something internal)' (Banfield, 1974, p. 281).

The main overall message of the book is therefore that it is detrimental for government to interfere in natural processes. These processes are those of economic growth, demographic change and the aspiration towards improvement and middle-class status. If these processes are allowed to operate then the problems will disappear (1974, p. 281). These processes have to be accompanied by the realisation that individuals are responsible for their actions and that improvement has to come from their own effort rather than sitting back and blaming society. Reformers should cease to reinforce the view that society causes problems and that therefore society must try to remedy them because 'by seeking a heavenly city, they may produce a hellish one' (Lampman, 1971). For Banfield, then, there seems to be no role for planners with their interventionist approach.

Siegan and the Houston experience

Proposals to reduce the amount of planning in this country often refer to the American example. It is sometimes said that the system there is more clearly oriented towards the protection of property rights and therefore carries greater clarity than the British system which relies on the troublesome definition of 'the public interest' (see for example, Delafons, 1969, p. 113). It has already been seen how antagonistic the New Right is to concepts such as the 'public interest'. However, the British advocates of the American system don't simply draw upon that system as it stands but refer to those American writers who consider that even the more limited American system needs pruning. One of the most quoted authors is Bernard Siegan (see for example, Walters *et al.*, 1974, pp. xi, 27, 83–4, 89–91; Adam Smith Institute, 1983, pp. 39–41) who advocates the removal of zoning regulations in the USA on the basis of a study of Houston. Siegan's general conclusion is that cities are better places if government ceases to interfere in the natural processes of the land market and that the example of Houston proves the point. He seeks the 'elimination of most government powers over land use' and believes that it is better to 'allow the real estate market greater opportunity to satisfy the needs and desires of its consumers' (1972, p. 1). Siegan applies this view to zoning powers and concludes that 'zoning has been a failure and should be eliminated: Governmental control over land use through zoning has been unworkable, inequitable and a serious impediment to the operation of the real estate market and the satisfaction of its consumers' and 'it is not even necessary for the maintenance of property values' (1972, p. 247). Four years later he is repeating the same message but applying it rather more broadly to the whole planning process. 'Public planning of land use is erratic, chaotic and irrational; it produces many more problems than it solves' and 'is doomed to failure in a representative society' (1976, p. 1). How does he substantiate this view?

He first criticises the planning process for lack of rationality; for being subject to political and popular pressures. He refers to the impossible complexity of issues that planners need to take into account in assessing the 'correct' use for every plot of land. In the end there is no rational way of doing this with the result that 'planning is unquestionably highly subjective, lacking those standards and measurements that are the requisites of a scientific discipline' (1976, p. 2). In his view this means that too much discretion is left with the planners, who end up making decisions based upon ideological orientation. Given the subjective nature of the decision-making process, political pressures, views and influence become of paramount importance. Planners are employed by politicians and will therefore be appointed for their ideological compatibility. At the end of the day 'zoning and other land use regulations are and have to be a tool more of politics than of planning' (1976, p. 3). The main reason for this is that 'there is simply too much money and power at stake in land usage for the planning process to be completely removed from political pressures' (1972, p. 8). However, the pressures not only come from the politicians but also come from the public. According to Siegan because public participation is an integral part of the zoning process with plenty of opportunity for public hearings, any 'ideal' plan will be subject to change as a result of the lobbying strength of different sections of the population. There is also considerable scope for graft and influence. Hence the planning process is cumbersome and subject to so many pressures that its outcome will be totally unpredictable and irrational.

Siegan contrasts this process with that of the market. The experts in the use of land are the developers and they have to risk their money and livelihood. They spend all their time considering these issues and gaining experience and have to be fully versed in the requirement of the consumer if they are to make profits or avoid bankruptcy. There is no such compulsion on the planner-politicians. They can develop their views and standards without this expert knowledge. Their only need is to ensure that they don't offend the politically powerful individuals and groups on whom they depend for votes. Therefore Siegan argues that the best 'planning' process is to allow the expert knowledge of developers to operate in conditions of unrestricted competition. This will ensure that customers' needs are fully met and also at the cheapest cost. Full competition resulting in maximum production will generate the best environment.

Siegan illustrates these views in relation to shopping development. He criticises planning interference in the provision of shopping centres claiming that it is impossible 'for anyone to determine the "correct" amount of business competition' (1976,

p. 133). Prevention of 'out of town' shopping centres is a block on progress. Under market conditions a new centre will have to provide something better to attract customers and everyone will benefit. If the old centres continue or adapt they will also provide a need; however, if they die it is because they no longer satisfy any modern requirement. In such cases it is the developer who is the expert and he should be left to make the decisions. According to Siegan it is therefore wrong that planners or sections of the public should interfere. 'If the experts in this field are the builders and the developers, why, paradoxically, does zoning require them to submit their proposals for final decision to the public and its representatives?' (1976, pp. 6–7). He describes as 'preposterous' the condition that was applied to a decentralised shopping development in San Diego that the final decision on the design and placement of buildings should be made by the city's planning commission of private citizens. He objects to such involvement because 'unfortunately, those with an axe to grind, local busybodies, and professional joiners have the most time for involvement' (1976, p. 7). He sees attempts to upgrade the proposed development as counterproductive as the developer will transfer the extra costs on to the consumer or cut back in other ways such as quality of materials, sound-proofing and so on. He believes that such interference is unwarranted and that development can be left to the self-regulatory controls of the expert developer having to satisy a market need and gain custom together with the finance institutions who will be checking things out before they risk their money.

According to Siegan another way in which the market is superior to planning is in reacting to innovation and change. This is a natural and inherent aspect of market process but one which planning finds impossible to deal with. Planning simply takes the past and projects it forwards, for example the policies towards shopping failed to take into account the arrival of the supermarket until after it had appeared. The lengthy and cumbersome procedures militate against the ability to be flexible and react to demands and innovation as they occur.

He also seeks to strengthen his argument by demonstrating the practical feasibility of doing away with zoning. For this he points to the success of the only large American city without zoning, Houston. His aim is to show that private control mechanisms have worked just as well if not better than zoning. He stresses that there have been several referenda on the issue of whether to have zoning and each time the popular reaction has been against the idea. Most of Siegan's attention is directed to restrictive covenants which are also referred to constantly by the British writers who draw upon Siegan's work.

They are seen as a good alternative to plans drawn up by local authorities. Although the covenants appeal to Siegan and others because they are private agreements it is interesting to note that the Houston planning authority has become very active in monitoring these covenants and ensuring their enforcement. The terms of the covenant are set out by the developer or financial institution prior to the sale of the housing. Afterwards any change in the covenants requires the approval of all the owners in the covenanted area. Often there are controls in the covenant that would not be covered by normal zoning procedure, such as architectural requirements, costs of construction, aesthetics and maintenance. When the period of the covenant expires then the controls over the land uses in the area cease but most covenants have an automatic extension provision.

Many covenanted areas have set up 'civic clubs' which are formed to act on behalf of everyone in the area and to ensure the enforcement of the covenant. Fees are collected that can then be used to pay the costs of any enforcement action and provide common facilities such as green areas, recreation facilities, insect control and even a police force. Although most restrictive covenants are applied to housing areas there are some examples in Houston of their application to industrial parks.

The main conclusion to be drawn from Siegan's work is that he believes in removing development control from the political process because all kinds of unsatisfactory influences operate producing uncertain, illogical and undesired results. Instead the city would be a better place to live and work in if the natural control process of the market and individual self-interest were allowed to operate. This means putting the control into people's hands in residential areas where voluntary agreements can be set up to control activity in the area and any nonconformity can be challenged in the courts using the restrictive covenant or laws of nuisance.

It is worth noting that the situation in Houston is rather more complex than Siegan and his supporters would have us believe. As the social and environmental problems of the city get worse the question of zoning is kept alive. Dissatisfaction has increased as the effects of the *laissez-faire* approach become more pronounced, for example traffic congestion, air and water pollution, toxic waste, flooding, poverty, and shortage of parkland. Feagin (1988) shows how the issue of zoning has been presented in ideological terms as a contest between individualism versus socialism with little regard for the practical results. However, the effects of a decline in the city's economic fortunes in the 1980s has forced a reluctant reappraisal of its approach to development. Several ordinances have been passed which provide limited controls over the form and location of certain developments. So if Houston is the image of the future there may yet be a role for planning.

The development of thought in Britain

Non-plan: an experiment in freedom

One of the first statements in Britain fundamentally to criticise the post-war planning system from a right-wing perspective appeared in *New Society* in 1969 (Banham *et al.*, 1969). It generated a lot of controversy and interest although it was not perhaps treated all that seriously by planners at the time. The authors drew heavily for their inspiration on the USA, quoting Jacobs and extolling the virtues of exciting free market American environments such as Las Vegas and Sunset Strip. The message of the article was indicated in its title 'Non-plan: an experiment in freedom'. According to the authors the concept of town planning had 'gone cockeyed' – no one knew any longer the purpose of planning – and they postulated that it was quite conceivable that society would be better without it. Their idea was to set up zones in which planning ceased to exist and which gave people back their freedom of action.

> The right approach is to take the plunge into heterogeneity: to seize on a few appropriate zones of the country, which are subject to a characteristic range of pressures, and use them as launchpads for Non-Plan. At the least, one would find out what people want; at the most, one might discover the hidden style of mid-20th century Britain.
>
> (Banham *et al.*, 1969, p. 436)

They attacked planners for creating a monotonous and predictable environment and for imposing their own values on others.

Then in a manner reminiscent of Banfield, the article ends with the words, 'Let's save our breath for genuine problems – like the poor who are increasingly with us. And let's Non-Plan at least some problems of planning into oblivion' (1969, p. 443). Now this article can be regarded as a bit of kite-flying but as shown in earlier chapters the wind of right-wing political ideology has strengthened to lift and sustain this kite.

The groundwork: the IEA and the Department of Land Economy, Cambridge

A look at the Centre for Policy Studies' Bibliography and Keith Joseph's reading list shows the importance of the Institute of

Economic Affairs (IEA) in influencing right-wing Conservative opinion. Another influence can be traced to the Department of Land Economy at the University of Cambridge whose Professor, D.R. Denman, chaired the Centre for Policy Studies' Land Policy Committee. Although there are minor differences between the authors who wrote under the auspices of these two institutions there is a large area of common ground, including a desire to review the town planning system radically. Pearce (in Pearce *et al.*, 1978) talks about the need to go back to first principles and question the exact nature of the problems that the market creates and whether planning is the best way of dealing with these problems. Are there better alternatives? D.R. Denman (1980) poses what he calls two radical questions: 'what is the purpose of the plan and what are the consequences of implementing it for private property rights in land?' (1980, p. 14). Meanwhile several years earlier the IEA produced a collection of essays which concluded that the question that practical men in government and industry should be addressing was 'not so much whether planning law has done harm or could be improved but whether the case for a planning law at all has been demonstrated' (Walters *et al.*, 1974, p.xi).

Having set the terms of reference at this very basic level one of the criticisms that is made of planning is its lack of a defined purpose. The argument presented is that planning arose in the particular conditions of war-time when there was an accepted and easily identified need to reconstruct the physical fabric of the country. However, as time has gone by this initial need has passed but the operation of the planning system has continued.

> What were new ideas, hope-filled and assertive in the war days and after have become commonplace and taken for granted today. Among them is town and country planning. We tinker with its devices and its mechanics and forget that it is a machine whose parts and assemblage were in large measure forged in wartime. It has served some purpose and is now creaky with age; even those who love it best are apprehensive of its future.
>
> (D.R. Denman, 1980, p. 37)

Denman points out that even though responding to the war-time needs, the 1947 planning legislation itself did not define the purpose of development plans. However, the argument is not just that the initial legislation was lacking in a stated intention; it is further argued that as time has passed this legislation has been altered and modified and extended with still no mention of the underlying reason. The result is a hotchpotch of a system built upon a basis of particular war-time needs that were never properly set out in the first place, have been subject to numerous adaptations over a long period of time and are no longer relevant. No assessment has been made of whether this resultant agglommeration satisfies current needs and there is still no statement of its purpose (Pearce *et al.*, 1978).

Although planning is criticised for not having a defined purpose it is also criticised for formulating policy in the name of vague concepts such as the 'community interest'. In a manner reminiscent of Hayek, D.R. Denman states that 'pursuit of social equality has proved a will o' the wisp for 300 years in capitalist and socialist countries alike' (1980, p. 11). He is particularly concerned that the pursuit of such vague concepts undermines property rights.

> In Britain we are in danger of lifting the community above the individual. Powers are given to ministers and officials to act in the name of the community with an arbitrariness which disregards the rule of the law and property rights of citizens – especially in land. Too many government policies are directed towards abstractions – planning, wealth distribution, environmental protection and so on.
>
> (D.R. Denman, 1980, pp. 3–4)

This tendency is seen by Denman to destroy the fundamental principle that government's paramount aim is to protect the individual's property rights.

Planning is also seen as ignoring the positive attributes of the market process and needlessly interfering in these processes. Mention is made of the pleasant and often planned environments that have been created in the past by allowing unfettered freedom of operation to the property market. Examples are given of Bournville, Saltaire, Port Sunlight, Edgbaston, Westminster, Bloomsbury, Bath, and Kensington (see for example, Pearce *et al.*, 1978, pp. 3–4; W.A. West, 1974, p. 25).

> Towns were built before tidy minds conceived 'town planning' supervised by 'town planners'. They were often more beautiful – and worthy of preservation – than towns or parts of towns 'planned' by officials and built by government: who shall compare parts of Bath or Lavenham with Council housing estates, New Towns, and the rest?
>
> (Walters *et al.*, 1974, p.xi)

It is suggested that a number of problems occur when planning interferes in the market and property rights. Many of these problems are to do with economic inefficiency but others involve land scarcity, coping with complexity and the creation of new power relationships. Planning lacks any monetary criteria that could lead to speedy and efficient decision-making. As a result delays and the imposition of

106

expensive conditions are common practice. The costs have to be borne by the developer and ultimately by the nation because 'decisions for the use of land, however made, will eventually affect the use and distribution of capital and labour in all their manifold forms' (D.R. Denman, 1980, p. 15). In addition there are the costs of actually administering the system. Planning is seen as part of the evergrowing bureaucracy which is a drag on economic growth, again creating costs for the nation. The cost of running the planning system detracts from the profitability of the market sector (see for example, Pearce *et al.*, 1978, p. 80; and Bracewell-Milnes, 1974, p. 95). These administrative costs must be borne in mind when assessing whether it is advisable to interfere in the market. It must be accepted that planning intervention will be at a considerable cost both to the developer and administratively. Therefore it may be preferable to allow the market to operate even if it is inefficient where the costs of this inefficiency are less than those of intervention.

Another criticism of planning, again reflecting the views of Hayek, is that it cannot deal with the complexity of society. 'To be able to control fully the vast array of interdependencies relating to urban land would seem to be almost impossible' (Pearce *et al.*, 1978, p. 82). On the other hand these interdependencies can be automatically taken into account in the market processes without the need for elaborate and time-consuming investigations and studies. Pearce believes that 'certain imperfections and injustices in the urban real property markets may have to be accepted (too), because the policy tools available cannot be sophisticated enough to ensure their control' (Pearce *et al.*, 1978, p. 82). The attempts to create these sophisticated mechanisms of control take up so much time that planning is unable to keep up with events and responds in an inadequate fashion. This hampers progress (cf. Hayek), 'because the collection, collation and dissemination of information is impossibly lengthy, virtually all statutory planning is years behind the times in outlook. As is well known, the market responds much more swiftly to changing circumstances if it is not impeded' (W.A. West, 1974, p. 31). Also it is asserted that planners' claims to be able to predict the future are false. They rely upon the controls they have and which allow some ability to direct future events. Without these controls they are no more able to judge what will happen than anyone else (Pennance, 1974, p. 13).

Planning is sometimes criticised for restricting the availability of land for development thus pushing up the price of land and hence buildings and housing. According to this view, if there were no planning restrictions there would be more land available and everyone would benefit from cheaper housing. This argument is particularly applied to the Green Belt where the restrictions are most severe and where the pressure of demand is high. The comment by Walters that 'the planning authorities have created an artificial scarcity of building land and so driven up the price to astronomical levels' (1974, p. 5) is typical of the viewpoint. However, this is not a view that is shared by all the commentators who would otherwise agree with the above criticisms of planning. For example Pennance (1974) states that inflation and encouragement to home ownership increase the demand for housing and rapidly push up prices and 'it is reasonably safe to assume that this effect would have occurred regardless of land availability and even with minimal or zero planning' (1974 p. 18). Similarly Goodchild lends support to this alternative view in arguing that when the demand for housing increases builders will seek more residential building land and the price of this land will rise: 'thus in general, high house prices lead to high land prices and not the reverse as is sometimes thought' (in Pearce *et al.*, 1978, p. 15). Without going into the detail of this debate it is sufficient for the purposes of this chapter to note that although scarcity of land for development is often used as one of the arguments against planning, there is not universal agreement amongst critics that restrictive planning causes higher land or house prices.

It was noted in the last chapter that one of the criticisms of state intervention (for example by Hayek and Friedman) was that it created a separated power structure. It displaced some of the conflicts and competition of the market-place into the realm of politics. This is also reflected in the right-wing comments on the planning system. The power to give or refuse planning permission is seen as open to corruption, and conflicts of the market-place are displaced into the political realm which is open to abuse. Famous cases such as Poulson are often evoked to show how bribery develops when large profits are at stake. 'Planning creates conditions in which each property-owner has an enormous incentive to wangle, cajole, threaten, use special influence and ultimately to bribe, and where politicians and civil servants have immense power and temptation placed in their hands' (Walters, 1974, p. 5).

As well as demonstrating the limitations of planning the writers from the IEA and Department of Land Economy, Cambridge, are also concerned to suggest changes to the system. The general objective that underlies many of the suggestions is greater freedom for the market and a planning system which is responsive to this market. The problem is that 'planners tend to regard as anathema the hotchpotch of apparently uncoordinated activities produced by freely operating markets' (Pennance, 1974, p. 12). This attitude has to change and Pearce describes how planners must gain a better understanding of market forces and therefore the effects of any planning actions. He advocates the greater acceptance of some of the undesirable characteristics of the market as these are inevitable in a mixed economy. As described above, attempts to over-control these shortcomings contain their own costs and uncertainties. A more considered use of intervention is required that utilises the 'benefits of a market based economy without destroying it' (for development of this argument see Pearce *et al.*, 1978, pp. 80–3).

How far, then, should planning try and cope with detrimental externalities created by the market? When can these be considered significant enough to justify intervention? It is difficult to find a precise answer to this question. For Pearce planning has an important role in this area; he believes that 'it would perhaps be the planners' single most important contribution to an efficient and just resource allocation if they were able to communicate the idea and importance of externalities' (Pearce *et al.*, 1978, p. 86). Pennance (1974) also considers that planning has a role in relation to externalities and suggests that more attention is needed in sorting out exactly what kinds of externalities require planning control. However, he also believes that there may be better ways of dealing with the problem. Attention needs to be directed to 'the establishment of precisely what are the important kinds of externalities generated by urban existence, how far they call for planning intervention and how far they might be handled by general modifications to our system of property rights' (1974, p. 19).

The problem of externalities therefore seems to be one area where planning is seen to have a role, though the extent of this role is unclear. Another area where planning is considered acceptable is that of conservation. For example it is said, 'there is one valid and beneficial function of planning control – conservation and preservation of our best heritage' (W.A. West, 1974, p. 28) and, similarly, that 'one single principle would seem to be sufficient to cover all the uses that ought to be controlled – the principle of conserving whatever is so beautiful, interesting or otherwise agreeable that there is a public interest in keeping it substantially unchanged' (Bracewell-Milnes, 1974, p. 92). However, although there is general agreement over the existence of controls for reasons of conservation, when it comes to how far this principle should be stretched then, again, differences of opinion appear. Green Belts are generally seen as a useful tool in aiding conservation but their application is considered to be too crude and as a result containing large areas with no conservation merits and therefore unnecessarily restricting the market in these areas (see S.E. Denman, 1974, pp. 56–7; Slough, 1974, p. 68).

The writers also give much attention to investigating alternatives to the planning system. Continual reference is made to the work of

Siegan for example in proposing voluntary agreements between property owners (Walters *et al.*, 1974, p.xii). These agreements might take the form of private restrictive covenants. A second area where planning could be replaced is where it tries to deal with neighbourhood conflicts. Instead of planning taking on the duty of protecting neighbours and either refusing permission or imposing conditions because of detrimental effects, the whole issue should be left to the courts and common law (W.A. West, 1974, p. 28). This is seen as preferable because judgements about what causes a detrimental effect are very subjective; eyesores, bad design of shop frontages and so on are not subject to abstract measurement. Pearce *et al.* (1978) describe how neighbours could get together and come to voluntary agreements about nuisance. If a developer imposes external costs on the others through his/her development then compensation payments could be negotiated.

In 1980, with these alternatives in mind, the Centre for Policy Studies' Land Policy Committee proposed a speedy review of the land-use control policy that has evolved since the Second World War. It was suggested that the review should cover the following issues:

(a) the purpose of public land use control;
(b) whether it could and should be detached from economic planning;
(c) whether its aims should be to safeguard declared and defined amenities and to manifest the cost and benefits of doing so;
(d) whether it would be possible or practical to reintroduce zoning control in a manner which would give greater freedom to the play of land market forces;
(e) to find ways and means of bringing the positive powers of decision making over land use which are inherent in property rights into the planning debate without passing the title to the land into public ownership;
(f) the extent to which attempts at public participation have unnecessarily clogged the planning machinery.

(D.R. Denman, 1980, p. 16)

At the same time the committee suggested that some reforms could take place at once. It suggested that structure plans could be abolished because of the unnecessary complication that they create. They could be replaced by a council's general purpose policy programmes. Then the local and action area plans should be made more responsive to market forces. This could be done through a discussion with the property holders in the area to see what development they would

be willing to undertake. The discussion could consider what induce-
ments or tax concessions would be required to persuade owners to
undertake development they would not otherwise have considered.
Pearce suggests a similar approach when he talks about the need for
'direct consultation with residents, traders, developers, and others
holding various rights in real property upon whose actions and good-
will the realisation of the plan depends' (Pearce *et al.*, 1978, p. 94).
He also advocates a 'key areas approach' which means that rather
than having a plan that covers all geographical areas, only specified
areas are selected for planning treatment and elsewhere good neighbour policies are adopted (1978, p. 92).

Thus, throughout the 1970s, a critical discussion of planning was
underway in Britain and the seeds of alternative approaches taking
root. The proposals were in tune with the ideology of the rising
faction within the Conservative Party.

Keeping up the pressure and the Adam Smith Institute

The 1979 election brought into power a government that was sympathetic to the ideas outlined in the last section. However, some
people still felt that it was necessary to keep up the pressure in order
to ensure that the policies reflected such a market-oriented philosophy. For example the newly formed Adam Smith Institute (ASI) set
up the Omega Project to create and develop new policy initiatives on
a wide range of government activity. They felt that this was necessary
because 'administrations entering office in democratic societies are
often aware of the problems they face, but lack a well-developed
range of policy options' (ASI, 1983, preface). Twenty working parties
were set up to introduce greater 'choice and enterprise' into government policies. Planners have generally dismissed the Adam Smith
Institute as an organization of cranks and extremists; however, analysts of Thatcherism have taken the influence of the Institute rather
more seriously and certainly many of the government's policy initiatives can be traced back to the work of the Institute. As Levitas says,
'it should not be supposed that the proposals contained in the
Omega reports are unlikely to be implemented, since there are connections between the ASI's organizers, the project's authors, and
government' (1986a, p. 82).

However, before examining the output of the working party that
looked at planning, brief mention will be made of another document
that sought to influence the government. This was called *New Life for
Old Cities* and was written by Anthony Steen and published by the
Aims of Industry. It was written after the riots in 1981 and sought to
ensure that government rejected intervention as a solution to the

inner city problem. The message is that 'fundamental reappraisal of the interventionist role of government' (1981, p. 5) is needed which accepts that increasing sums of public money are no answer to the decline of the inner city. Instead private enterprise in the form of banks, insurance companies, pension funds, building societies, property companies, small and large firms, multinationals, voluntary bodies and community groups, should not only be consulted but also determine policy and its implementation.

In *New Life for Old Cities* planning is not given a role and indeed is seen as the major cause of inner city decline and the riots. Drawing on Jacobs, Steen concludes that

> the one way of bringing back safety to the streets of our cities is to provide new scope, new opportunites for them to become alive and busy again. There should be places where people will want to come, live, work and shop. This means abandoning a rigid zoning policy, doing away with all manner of planning restrictions and burying Structure Plans. Only in this way can new life be brought back into the streets and, with that life, the surveillance and mutual policing which leads to safety and to security. Yet, so long as the planners hold sway, the life force of our cities will continue to drain away.
>
> (Steen, 1981, pp. 62–3)

Planners are seen to contribute to the decline in the inner city through their policies of clearance and decentralisation, their rigid zoning and their imposition of standards.

According to Steen one of the results of this action by planners is that a lot of money is fruitlessly channelled into deprived areas and wasted rather than encouraging wealth creation. The market would change this as there would be no wastage and the inner city economy would re-establish itself. The key to this approach is to tap the entrepreneurial drive of people. This requires a whole change of attitude; a shift from a reliance on bureaucracy to one of encouraging innovation and enterprise. Steen asks, what are the characteristics of the entrepreneur who will save the inner city?

> He is an innovator, which means he is likely to be an individualist. He is 'Mr Eccentric rather than Mr Normal', and he is most unlikely to be 'Mr Committee-Man'. He is probably intelligent and capable but not in the usual way, and may be very difficult for 'Mr Bureaucrat' to recognize. A big funding quango or a huge integrated committee of planners, or councillors, may have nothing in common with him, and are very likely to pass him by.
>
> (Steen, 1981, p. 33)

So according to Steen it is necessary to abandon attempts to solve the problems through local authorities and bureaucracies and devise new organisations that will recognise the entrepreneurial spirit. These organisations will be private organisations of various sorts in which the work is done by staff seconded from the private sector such as banks or by local people themselves. Such organisations will aim to 'market the cities' assets' and attract funds and interests from industry, insurance, banks and other financial institutions. He suggests privatisation of much of the local authority service and contracting out work at present done by council staff, including architecture and planning.

The critique of planning undertaken through the auspices of the Adam Smith Institute can be found in the Omega Report on Local Government Policy which covers three areas of activity: finance, planning and housing. The planning section builds upon a pamphlet written a year earlier by Robert Jones (and published by the Institute) called *Town and Country Chaos* (1982). Once again the critical approach taken is to set out the faults of planning, to pose the question of whether planning is necessary and suggest alternative mechanisms. The ASI adopts the view, expressed by the writers reviewed in the previous section, that when considering the merits of planning it must be remembered that the planning process itself carries a burden of cost.

Perhaps the broadest level of critique adopted by the ASI is the rejection of the concept of community values. It is said that planning promotes these community values in opposition to the pursuit of personal profit and that this promotion is not compatible with a free society. One of the problems of the British tradition of planning is seen to be that it 'denies that the owners of property have any automatic right to develop their property to best advantage' (R. Jones, 1982, p. 4). In any case, it is claimed that the effort of planners to promote community values hasn't worked. The general public has had no influence over events as these have been controlled by the 'planning class' or pressure groups. Jones, in similar vein to Siegan, explains how the politicisation of planning has produced pressures of illicit payments and that

> even without this kind of corruption there is the more insidious kind in which those with influence on local politics are able to achieve more decisions in their favour than neutral consideration might provide. The presence of a large and vociferous group in the public gallery, for example, can greatly influence council decisions.
>
> (R. Jones, 1982, p. 11)

The Omega Report's use of the term 'planning class' carries the implication that planners are isolated from the rest of society and
impose their views upon decision-making. In particular they look at
situations with their middle-class values to the detriment of working-
class and poor people. Examples are given of the planners' aim to
ensure 'town identity' which may lead to conservation rather than increased convenience and provision of facilities. This imposition of
values is also said to occur in the segregation of land uses through
zoning which is based upon planning ideals, 'but the ideals of the
planning class do not always reflect the wishes of those who live
under their rule, particularly those of poorer people, who have been
moved far away from the shops and workplaces they depend on' (ASI,
1983, p. 33).

Another problem seen as inherent in the planning process is its
unpredictability. There are two aspects to this argument. First, planning is criticised because its aims are very indistinct and open to
interpretation, for example what exactly is meant by 'local character'
or 'intolerable mixture of uses'. Second, there is a large amount of administrative discretion possible resulting in great variation from
place to place and planner to planner. Instead of all this uncertainty
and discretion it is suggested, with reference to Hayek, that a generally applicable set of rules and regulations, known in advance, is what
is required. So planners generally are seen to impose their own
values bogusly in the name of the 'community interest' and to do so
in an arbitary fashion. Not only this but also their activities are a great
cost to society. There is first the cost of preventing wealth-creating
activity which could benefit the economy. This happens when new
enterprises, and in particular small businesses, are prevented from
occurring in the name of some abstract concept of 'order' or 'nonconformity'. Second, great costs are generated because of the lengthy
processes, expensive delays and possible appeal costs. These costs are
borne by the developer and lead to higher prices. It is also claimed
that the unnecessarily severe standards imposed by planners, for
example over density or quality of materials, increase building costs
and are better controlled by other means.

At the end of the day it is claimed that planners have not, even
with the financial and political costs involved, succeeded in their aim
of improving the environment. The uniformity created by strict control and the poor state of council estates are given as examples of this
failure. So planners should cease trying to pursue these aims through
the control of land use. There are other instruments of control which
do not carry the same problems and planners should concentrate on
trying to predict future needs and help the market meet these, for
example, through the provision of infrastructure. This does not imply

that society can do away with some of the other activities currently pursued by planning, for example control of nuisance, conservation or maintaining standards, but that these functions are better performed in other ways.

The Omega Report sets out its reforms to the planning system which it thinks should be dismantled and replaced in the longer term. It states that 'the aim of these reforms is to replace the present planning policy and procedure by a system which retains protection for historic and rural areas, but which otherwise generally assumes planning consent' (ASI, 1983, p. 45). However, it accepts that the present planning system cannot be dismantled overnight and proposes that a number of experiments should be conducted to test out the alternatives.

The dismantling of the planning system is based upon the view that there are other control mechanisms that would work better. Four mechanisms are mentioned: economic forces, the laws of nuisance, central regulation and private institutional controls. Economic forces will result in land being allocated to the most appropriate uses without the need for planners' intervention. Thus, for example, economics will prevent a petrol station in a quiet cul-de-sac or large industry in a quiet residential neighbourhood. A strengthened law of nuisance and its enforcement would ensure that planners do not need to concern themselves with the problem of 'externalities' or neighbour conflicts. It is suggested that the conservation role of planning could be undertaken centrally by the Department of Environment (DoE) and thus avoid local discretion and ambiguity. The DoE would administer historic buildings, conservation areas, Areas of Outstanding Natural Beauty (AONBs), Green Belts and listed buildings. Such controls would mean that 'the much feared hamburger-stand-in-the-middle-of-feudal-England problem' (ASI, 1983, p. 40) would still be prevented. Finally, drawing very heavily on Houston, a number of private forms of control are suggested that could ultimately replace the public planning system.

These private forms of control are tribunals, private covenants and private codes of practice. 'The major structural reform which suggests itself is the replacement of the entire system of detailed planning procedures by a series of low cost and easy-to-use land use tribunals' (ASI, 1983, p. 42). In other words the development control system would be replaced by a legal process based on the law of nuisance. Widely published 'tolerance guidelines' could be produced by the DoE to inform developers of the standards required. Private covenants could be encouraged to cover residential areas, as in Houston. Insurance underwriters would check on standards of buildings through the premium mechanism and unsound buildings would be faced with high compulsory premiums. Although this suggestion is presented in the Omega Report as a replacement of planning controls it has greater relevance to the system of building regulations.

Having set out these longer-term proposals the report then turns to ideas about immediate changes. It mentions the need to reduce delays and suggests that some work could be contracted out to consultants, planning fees could be refunded or compensation given for delays and enquiry decisions could be made at the end of proceedings. It proposes that costs of appeals should not be borne by the applicant. In the interest of promoting small business, it also suggests that there should be no control over using a home for other purposes. Only if a complaint is received should it become a matter requiring attention.

Lastly the report suggests that a three-fold zoning system could be introduced in the immediate future which would greatly simplify the planning system and free many areas from unnecessary restriction. The first zone, 'restricted zones', would still be controlled although the procedures would be simplified. Such areas would be administered by the DoE and cover the conservation-type areas mentioned above. There would then be 'industrial zones' in inner city and derelict areas in which the only regulation would be 'safety, public health, pollution and other nuisance control' (ASI, 1983, p. 44). The rest of the country could be covered by 'general zones'. The report does not make it clear how the treatment of these last areas would differ from 'industrial zones' but it might be assumed that they occupy some middle ground between strict conservation regulations and the planning-free 'industrial zones'.

The work of the Adam Smith Institute therefore seems to repeat many of the earlier criticisms of planning and draws heavily upon ideas from the USA, particularly Houston. However, it does go further than previous critiques in its strong promotion of a dismantling of the planning system in the longer term. It also extends slightly the debates on the possible alternatives. It is not clear however how the short-term suggestions relate to the longer-term ones, how one assesses the success of experimentation or decides when to move into the dismantling phase.

Conclusions

There are a number of common themes that run through the contributions reviewed in this chapter. The critique of the current planning system perhaps carries greater consensus than the suggestions on how it should be modified or replaced. These ideas on possible alternatives, though pursuing a common direction, show much variation in detail. In general terms there seems to be agreement that the planning system has enormous faults and that the process of determining the use and control over land should be shifted away from the planning system, and its political context, into the market and legal arena. In other words the market processes should be paramount, backed up by a strong legal framework. Such a change is specifically linked by Siegan, Banfield, Steen and the Omega Report to the need for a greater emphasis in society generally on the free market and individualism. This shift from a political arena to a market and legal arena would enable generalised and impartial rules to be imposed. These rules would replace the uncertain and ambiguous decisions of planners which vary from place to place and person to person.

The critique of planning is undertaken at both a very broad and a detailed level. Perhaps the broadest comment is that of Banfield (1974) and Banham *et al.* (1969) in which they claim that planning deals with a non-problem. The real problems of cities are said to be focused on poverty and outside the remit of planning. These problems are caused by individuals and their attitudes and are thus seen as best dealt with through an attack on these attitudes or through government tax policies. Not only are planners irrelevant but also they actually aggravate the problem of urban poverty. According to Jacobs and Steen the decline of inner cities can be blamed upon past planning policies.

However, according to most of the authors reviewed in this chapter, the main problem created by planning is that it interferes with the market thus preventing it from fulfilling its proper and beneficial function. Both Siegan and the Omega Report point out that much of the land use allocations resulting from planning activity would happen anyway if the market was allowed to operate freely. Most writers talk about the complexity of cities and the myriad factors and interconnections that planners need to consider in making decisions about land use. They believe that their efforts are bound to fail because understanding such complexity is an impossible task. The result is a false simplicity that is bound to have adverse side effects.

Almost all the writers mention the detrimental impact that planning has on wealth creation. This is a cost of planning that has to be borne by society. Thus Banfield and Jacobs conclude that a dynamic and prosperous city economy requires inefficiency in its structure and land use and the rational pursuit of order by planners and others will kill off the potential economic growth sector, innovation and experimentation. Banfield goes as far as saying that the ugliness of American cities is a 'non-problem' and implies that a 'city beautiful' approach leads to fossilisation. This broad theme of the dynamism of the market is picked up at a more detailed level by Siegan and Steen

117

when they promote the role of the developer and entrepreneur. According to them these people are the experts in land use and should be left to the job whether pursuing their own development schemes or devising solutions to the inner city problem. Historical examples of pleasant private estate developments are referred to by a number of authors.

Are there any problems that the market creates or cannot solve? There seems to be both variety and uncertainty in the response to this issue. Pearce and Pennance believe that there is a role for planners in dealing with externalities although they seek a clearer definition of this role and a greater understanding by planners of the market process. Jacobs seems to stand alone in asking planners to control the detrimental actions of those institutions such as banks and insurance companies that have the money to dominate the land market creating adverse affects on diversity. Experience in Houston shows that even in this 'Mecca' planning controls to overcome externalities are still needed. On the other hand the Omega Report believes that the whole issue of externalities can, in the longer term, be taken out of the hands of planners and dealt with by legal processes. Although other writers may not go as far as this there is widespread support for more use of the laws of nuisance.

Alongside the promotion of the market as the best decision-making mechanism, there is the attack on the political decision-making process and the associated administrative and professional procedures and attitudes. Several authors make reference to the way in which the political process can be influenced in a corrupt way. There is also an antagonism, for example as shown by Siegan and Jones, towards the way in which participatory opportunites are used by groups or individuals to try and influence results. The Land Policy Committee of the Centre for Policy Studies suggests, through Denman, that the public participation aspects of the planning process need to be re-examined. On the other hand Jacobs advocates greater involvement of local people who know about the details of their areas.

Several writers, for example Banham *et al.*, Jacobs, and Banfield, complain about the way in which professional planners impose their own values on communities. The Omega Report claims that this is done to the detriment of the poorer classes. It is furthermore suggested that planners impose these values under the guise of vague and meaningless concepts. Denman also refers to these useless concepts, mentioning 'social equality', 'wealth distribution', 'environmental protection' and 'planning' itself. The fear is that these concepts are used to undermine the principle of property rights. In similar vein Jacobs criticises the simplistic application of concepts, such as 'self-containment', 'decentralisation' and 'neighbourhood', and complains about the antagonism to the mixed use of land. Thus the picture being presented is that planners are operating within a corrupt and irrational political system which is easily influenced by strong-minded groups or individuals. At the same time planners are imposing their own values and justifying their actions by evoking a number of meaningless concepts without having any clear sense of purpose. However, the overall effect of planners' attitudes and approach is to support the principle of the interventionist state and is antagonistic to the market. Banfield refers to middle-class government officials who promote and protect government intervention as a solution to city problems. Such an attitude is based on the ethic of 'service' and has the disadvantage that it blames society for problems rather than 'individuals'. This ethic detracts from 'self-help' which is seen by New Right authors as the basis of a better solution.

Then there is the problem of the costs that arise from the bureaucratic and administrative processes involved. Writers refer to the costs that the process generates for the developer (and ultimately society as these costs are passed on). The main target of these complaints is the time taken to process applications. Mechanisms for overcoming delay are seen by the Omega Report and many of the IEA writers as the most important short-term improvement. Some of these writers have also pointed to the 'cost' created by planning in causing higher land and house prices when land supply is regulated. Contracting out planning functions is seen as another way of reducing costs.

So what roles remain for planning and what alternative mechanisms, other than the free market, are suggested? Some of the writers see planning as having a role in relation to market externalities. Conservation is another role that is sometimes mentioned but it is not clear how far this role should be extended. Green Belts, AONBs, conservation areas and listed buildings are mentioned by the Omega Report and the IEA writers. However, the Omega Report also suggests that the function should be centrally administered to avoid discretion and disparity between areas. The Omega Report and Siegan, drawing on the Houston example, suggest that planning could also have a predictive role in providing information on likely future demands to help the private sector's investment and also to ensure the required infrastructure. The overall position of planning, particularly as presented by the Omega Report, is a minimal one operating in selected areas and, in general, allowing the market to operate freely. Several authors, for example Pearce, Banham *et al.* and the Omega Report, suggest that planning should be more selectively applied. As a result many areas would be free from any planning restrictions, while others, for example conservation areas, would operate under a strict regime. Denman and Steen both suggest scrapping structure plans and that the remaining local plans should either be more sensitive to the market or actually devised by the private sector.

The use of the laws of nuisance has already been mentioned. This forms part of a broader theme of shifting many planning controls into the legal arena. Thus the Omega Report suggests that planning restrictions on the use of houses for non-residential purpose should be lifted, and action taken only if there are complaints. This approach is expanded into the idea of the land tribunals for dealing with neighbour disputes. Replacing planning controls by private covenants is advocated by a number of writers including many of those from the IEA, Siegan and the Omega Report. These covenants would mainly be applicable to areas that were predominantly residential and would again rely considerably upon the legal process for their formulation and enforcement.

Finally, the ideas explored in this chapter will be linked to the framework set out at the end of the last chapter. To what extent do these writers on planning reflect the shift in emphasis concerning the principles of decision-making? It is clear that there is a distinct preference for market criteria in all the work covered in the chapter. Hayek's argument about the complexity of society is taken up by the writers on planning who refer to the over-simplification produced by planners. Attempts by planners to interfere in market processes create costs and affect the wealth-creating potential of society. The economies of cities are detrimentally distorted and developers' costs increased. The attack by Hayek and Joseph on the meaningless concepts that are employed as a substitute for the criteria of the market is strongly mirrored in the planning literature. A whole list of vague terms used by planners is identified for criticism, as indeed is planning's lack of any clear overall purpose. Such terms are seen to justify planning intervention and form a threat to the principles of both the market and property rights. However, there is a rider to this critical stance. It is accepted that some kind of intervention in the market and property rights is necessary because of the 'externality' or 'bad neighbour' effects, although there are different views over the necessary extent of this role.

There is rather less coverage in the planning literature reviewed of the procedures of decision-making. However, several authors, following Friedman and Hayek, discuss the way in which corruption and bias can enter the system because planning is in the political arena. This leads to suggestions that there should be less participation and democratic involvement and more reliance on the entrepreneur. This criticism of planning also leads to the recommendation that planning decisions should be removed from the political arena not only to the market but also to the legal system. Thus suggestions are made for the greater use of the laws of nuisance, tribunals and restrictive covenants.

Many of the author's comments on the role of planners reflect the 'anti-bureaucratic sentiments' identified in earlier chapters. Banfield's idea that planners detract from the principle of self-help and individual responsibility can be compared with the views of writers such as Schumpeter while the Omega Report's 'planning class' can be clearly seen as a sub-species of Kristol's 'new class'. Hayek's desire to identify 'rules of law' can be linked to the statements of Siegan and others to create a general framework of laws, cf. Houston, that removes opportunities for administrative discretion.

The writers on planning outlined in this chapter reflect to a large degree the approach and philosophy of the New Right identified in earlier chapters. To what extent do these attitudes lead to a reorientation of planning along the lines of the reorientation of the state described earlier? The writers have far less to say on this matter although, in general terms, a more restricted and selective planning system is envisaged. It is suggested that planning is still needed to provide information and infrastructure for the market, to carry out conservation in selective areas and deal with certain 'externalities' of the market.

The next stage of the investigation is to explore how far the ideas discussed in this chapter have been pursued in practice, as evidenced in changes to the planning legislation. Have pragmatic pressures led to deviations from the principles described here? As outlined in the introduction, this investigation of practice will be explored under three headings; modifications to the planning system, mechanisms that by-pass the system and initiatives that could potentially replace the system.

Chapter six

Modifications to the planning system:
1   Development plans

This chapter examines those views and actions of the government since 1979 which relate to the modification of development plan legislation. In the same way as the New Right built on earlier ideas, so the agenda for changes to planning in 1979 picked up on the work already in progress. The Conservative Party in opposition in the 1970s had been developing their approach to planning through two Policy Groups chaired by the opposition spokesman, Hugh Rossi. These Groups reported that the planning system needed to maintain a balance between the public, local authorities and developers and that there was a need to restore this balance as the needs of developers were not being adequately met. The reports contained many piecemeal changes to the planning system, particularly to development control, and many of these changes were included in the early programme of the Thatcher government. However, as Rydin (1986) points out, the tenor changed once the new government was in office. There was a shift from a stress on the need for balance in the planning system to a stress on the need for minimal public intervention.

This more strident tone was reflected in the speeches of the various Secretaries of State from 1979 onwards and there has been considerable consistency in their approach. This is illustrated in the speeches made to the Royal Town Planning Institute (RTPI) Summer Schools. The approach has been to reaffirm the importance and value of the planning system but then to proceed to outline its considerable deficiencies and need for reform. In 1979 Heseltine said, 'I have no intention of wrecking the planning system developed in the last 40 years or so in this country' (1979, p. 25). Two years later he repeated his support of the planning system when he said:

> I remain, as I have long been, committed to the concept of planning. Britain would be the poorer without it and it is to me unthinkable that the broad philosophy of development control

will ever be set aside. Our land resource is too limited, the pressures on it too great to contemplate such a prospect.

(Heseltine, 1982, p. 11)

In 1983 it was the turn of Patrick Jenkin to state that:

We understand and support the positive contribution that the
planning system properly administered makes to our national life.
Our land resources are so few and the pressures on them so great,
that to abandon planning is unimaginable. The Government has
maintained and will continue to maintain the main thrust of the
planning system.

(Jenkin, 1984, p. 15)

More recently Lord Elton took 'a positive view of town and country
planning' (1986a) and, in 1986, Nicholas Ridley continued to express
this commitment to planning when he concluded:

There will be those who say that we are intent on weakening the
planning system. Nothing could be further from the truth. A strong
and effective planning system is the best way to encourage sensible
development and to protect the countryside where that is necessary.

(Ridley, 1987, p. 41)

However, as mentioned above, these statements were also accompanied by proposals to change the system. The usual reason given for
these proposals was the need to ensure that the system adapted to the
contemporary attitudes and needs of the population. 'As our society
changes, so the planning system must change too, in order to retain
the confidence on which it depends. Changes in industry, in population and in public attitudes require a response from the planning
system' (Jenkin, 1984, p. 15).

The purpose of this chapter is to examine those proposals that
have led to modifications to the development plan system, to analyse
the significance of these modifications and explore whether they
demonstrate any consistency, reflecting the ideological themes identified in earlier chapters. Could it be said that these changes rather
than simply keeping the system 'in tune with today's needs' (Elton,
1986a) are pursuing a particular and consistent approach that is in
tune with the ideological priorities of Thatcherism? The first indication of the government's attitude came with Heseltine's changes to
submitted structure plans. The main issues raised by these changes
and their embodiment in the Local Government Planning and Land
Act of 1980 are examined. Attention then turns to the local plan
level, including the Unitary Development Plan initiative. Finally,
ideas for reorganizing the development plan system are reviewed.

The first attack on structure plans

The first clear statement of the new government's approach to planning came in the Secretary of State's reports on the structure plans that were awaiting central government approval. Ministers were putting pressure on those authorities that had not submitted their plans to do so quickly. Ministers were also expressing the view that the process should be speeded up and that one way of doing this was to reduce the scope of the plans. As Heseltine (1979) expressed it, 'structure plans are not a vehicle for displaying every conceivable matter of interest to a county council'. The ministerial reaction to a number of these plans is reviewed in order to identify the direction of central government thinking in these early years.

South Yorkshire was the first structure plan to receive attention. The DoE said that numerous modifications were necessary to bring the plan into line with the economic changes that had occurred since the plan was prepared. These modifications included alterations to the proposed restrictions on car parking, shopping and office developments and the plan's recommendation that future resources should be concentrated on the deprived Dearne Valley. In disagreeing with this last proposal Heseltine stated that the 'greatest opportunity for increasing the prosperity of all the people of South Yorkshire, including the Dearne Valley, rests on encouraging the natural economic growth of Greater Barnsley, Greater Doncaster, Greater Rotherham and Sheffield' (*Planning*, 1979). Jowell and Noble (1980) have pointed out how this last modification prevents the county developing policies in the structure plan that use redistribution of resources as a tool of social policy. They quote the Notice of Approval which states 'that structure plans should not include policies which would tend to inhibit the growth of major centres with the greatest natural potential for new industrial, commercial and office development' (1980, p. 305). Thus Heseltine is stating quite clearly that plans cannot try and modify market forces, either through restrictions or through the identification of areas for positive discrimination (see also Darke, 1979, for the ideological conflicts that took place at the Examination in Public).

The Norfolk Structure Plan followed and again Heseltine requested numerous modifications. An important aspect this time was his disagreement with the county's attempt to control population increase and limit development because of the costs of infrastructure provision, particularly education. He raised the population increase proposed from 3,500 people per annum to 4,000–5,000 and deleted the policy on educational capacity. Jowell and Noble (1980) point out that this is an example of the Secretary of State restricting the plan's scope to relevant 'land-use issues'. A number of plans at this time tried to introduce the concept of 'essential local needs' to prevent certain areas becoming dominated by commuting populations with the resultant loss of local facilities and increase in house prices. Jowell and Noble report how the Secretary of State took objection to this kind of policy. However, in the case of the Norfolk Structure Plan the protection of 'local needs' was expressed as part of the reasoned justification rather than in the plan policies and as a result escaped the Secretary of State's objection.

Then, in April 1980, Heseltine approved the Central Berkshire Plan on condition that land be found for a further 8,000 houses. As Bruton (1983) points out he seemed clearly set on reducing restrictions on the operation of the housebuilding industry. Later in the year major modifications were again made when the report on the Manchester Plan was released. This deleted more than 40 per cent of the policies and a further 20 per cent were substantially modified. This action led to the comment that the structure plan approval system was creating increased centralisation in which the DoE was imposing its view of acceptable planning content (Bruce, 1980). Again in February 1981 the Secretary of State announced extensive modifications, this time to the Tyne and Wear Structure Plan. The modifications covered all the major policy areas and often overruled the panel's recommendations. Changes required lifting restrictions on shopping developments in certain areas and increasing land for housing by 7,000 dwellings.

These modifications in the first two years of the government's life illustrate a number of themes:

1 The willingness of central government to get involved in the content of the plans as shown by the extensive nature of the changes.
2 The intention to free up the restrictions in the plans and allow market forces to operate, both in terms of housing and economic growth in 'natural areas'.
3 The limitation of content to appropriate planning matters and the removal of social policies. As Jowell and Noble (1980) show, the Secretary of State requires that social aspects are restricted to broad statements of aims or the 'reasoned justification' section. Then, as the 'reasoned justification' is down-graded in importance the social dimension of the plans is further diluted. (In the Town and Country Planning (Structure and Local Plan) Amendment Regulation, 1979, the 'reasoned justification' became a non-statutory part of the plan.)

These themes are well summarised by Jowell (1983). In commenting on the role of the Secretary of State in monitoring structure plan policies he concludes that:

> underlying the policy debate is a political process largely camouflaged by the neutral language of professional planning. The approval process permits a high degree of central control over local authorities. The Secretary of State employs that control to strike down not only social and economic policies in general, but particular policies of which he disapproves. Occasionally he pursues policies of his own that have not been tested in any public forum.
> (Jowell, 1983, p. 45)

The Local Government, Planning and Land Act 1980

The Local Government, Planning and Land Act 1980 was the first major item of legislation affecting planning passed by the 1979 administration. The Act covered a wide range of issues including the setting up of Urban Development Corporations and Enterprise Zones, which are covered in later chapters. Of relevance to this chapter are those sections that directly modified the planning system.

Under the Act (Part 9) development control powers are the responsibility of the district authority who have to consult the county only in certain circumstances. Exceptions to this are mineral extraction and waste disposal, where the county retains its powers. The effect of this part of the Act is to make County Councils impotent in dealing with matters of strategic importance and unable to ensure the implementation of policies and proposals included in structure plans. The previous relationship in which the strategic authority could 'direct' a local authority decision if it didn't conform with the structure plans is replaced by the much weaker requirement that the local authority 'consults' with the strategic authority. This change led Wilbraham (1982) to comment at the time that the Local Government, Planning and Land Act 1980 looked like the first step in transferring all power from counties to districts. Districts become responsible for deciding whether a particular application constitutes a departure from the structure plan. According to a survey conducted in 1983 by the AMA, fundamental departures from the structure plans have occurred in a large number of cases since the consultation procedure of the 1980 Act has been put into practice (Greater London Council, 1985).

A second area of change brought about by the Act concerns the relationship between structure and local plans. Since the inception of this two-tier system it had always been expected that local plans would follow on after the structure plans had been prepared, thus ensuring that they conformed to the strategic context. This principle was relaxed in certain specific inner city areas by the Inner Areas Act of 1978. However, the Local Government, Planning and Land Act of 1980 allowed local plans in all areas to be prepared and adopted without waiting for an approved structure plan. This again reduced the counties' ability to ensure the implementation of strategic policy.

A third change included in the Act sought to speed up the plan-making process. Thus sections 88 and 89 provide for a speeding up of the preparation of both structure and local plans. The need for surveys is reduced and the participation part of the process diminished in time and scope.

The changes introduced by the Act are elaborated in Circular 23/81 and Circular 22/84 (DoE, 1981c, 1984b) which draw together the Regulations and incorporate them in the new Memorandum on Structure and Local Plans, replacing that of 1979. These Circulars reinforce the points made above. Speed and simplification of the process are key objectives leading to the removal or reduction of certain stages in the process of plan production. Continual reference is made to 'keeping down the cost' and the need to 'concentrate on essentials' (DoE, 1981c, p. 1). Structure plans are criticised for over-elaboration in their studies and surveys and also in their policies. In any review of structure plans the work programme has to be discussed with the regional offices of the DoE to eliminate unnecessary work and prevent large demands on staff resources. Rather than the previously suggested five-yearly review a much more restricted approach is required with the need to review only when major changes demand it. A report of survey for structure plan review is no longer necessary. It is assumed that in many cases survey work will also be unnecessary for local plans as they can rely on the structure plan research. In many cases a local plan may not be needed at all as the structure plan would be considered an adequate framework or there may be little pressure for development. As Circular 23/81 puts it:

> The objective will be to make sure that there is a clear need before preparing a local plan and that work proposed on the preparation of the local plan to enable it to be placed on deposit for objection is realistic, justified and can be carried out within about a year.
>
> (DoE, 1981c, p. 3)

Participation is also pruned in the interest of greater speed. First there is no requirement to undertake publicity or carry out consultation at the survey stage and participation before the formal representation stage should be confined to one occasion. The period of six weeks, previously regarded as the minimum for representations on the draft plan, is to be regarded as the norm. Longer periods for representation and any additional participation opportunities, such as previously adopted by authorities, have to be clearly justified as necessary (DoE, 1981c, p. 4). Under the new regulation the Secretary of State can decide that a decision can sometimes be reached on a structure plan without an Examination in Public and similarly a local authority can dispense with its Local Plan Inquiry if no objectors wish to appear at the inquiry.

As indicated above there has been a tendency in the Secretary of State's modifications to structure plans to confine planning to land-use matters and to reject social policies. As pointed out by Bruton and Nicholson (1987) this tendency has been consolidated in the new Memorandum. First it confirms that the written justification of the plan is no longer part of the statutory document thereby reducing the legal status of this part of the plan. Then the social policies are excluded from the statutory policy statement which has to confine itself to land-use aspects. Quoting the Memorandum:

> Non land use matters, for example, financial support, consultation arrangements and proposed methods of implementation should not be included as policies or proposals in structure or local plans.
>
> (DoE, 1984b, p. 41)

Then again in relation to housing policies:

> It is inappropriate to include in structure or local plans policies or proposals which might lead to the assessment of planning applications on the basis of the identity, personal needs or characteristics of individuals.
>
> (DoE, 1984b, p. 44)

The Circular also makes it clear that the policies should not specify areas for positive discrimination, for example through the use of special facilities or grants.

The preparation of non-statutory local plans which in the past have exceeded purely land-use policies are also to cease. Bruton and Nicholson (1987) conclude that

> the restricted content of local and structure plans, allied with the prohibition of non-statutory local plans if adhered to by the planning authorities would ensure that plans produced within the town and country planning system would be simply land use allocation proposals.
>
> (Bruton and Nicholson, 1987, p. 41)

General reviews of the Local Government, Planning and Land Act, which also encompass its financial aspects, stress the increase in centralisation created by the various measures (Raine in Raine *et al.*, 1980; McAuslan, 1981). Loughlin (1981) concludes that the Act also demonstrates the way in which the government is seeking to replace Welfare State ideals with 'value for money' and make local government more responsive to the wishes of central government and the private sector. The general conclusions that can be drawn from the planning aspects of the Act and the subsequent Circulars are a downgrading of the powers for strategic planning and the ability to implement this level of policy, a restriction in the scope of planning towards strict land-use matters and a reduction in the opportunities for public participation as the procedures become more streamlined and centralised.

Local plans and conflict in the City

The Secretary of State has considerable control over structure plans through ministerial approval and modification. This degree of control does not exist over local plans although the Secretary of State can always use his 'last resort' call-in powers and local plans must conform to structure plans where these exist. However, it is possible for the government to apply its ideology indirectly through other agencies that share its outlook. These agencies can become objectors at local plan inquiries and the inspector then has to consider their views. Two examples which illustrate this approach are considered: the City of London Plan and the North Southwark Plan. The examples have been chosen to highlight another way in which ideology has impinged on the planning system rather than to be representative of local planning since 1979 (for a full account of local plans in Britain see Healey, 1983 or Bruton and Nicholson, 1987). In the City of London case the objection came from the Policy Studies Institute and in Southwark from the London Docklands Development Corporation. As will be seen, the examples also illustrate conflicts over the purpose of planning. In Southwark the conflict was between planning to aid development and planning to satisfy the needs of the local community (see Thompson, 1987). The Southwark case will be dealt with in detail in Chapter Eight when considering the Urban Development Corporation initiative.

The City of London produced a draft local plan in November 1984. Although this plan stated the importance of the City as a financial centre it also placed great emphasis on the conservation of the physical environment. The general philosophy of the plan was described in the following way:

The basis of the Plan's strategy must be to resolve the conflicts

between the pressures for development necessary to the commercial well-being and the need to protect its special character,
historic heritage and the range of facilities and activities which
make the City an attractive place in which to work and to visit.

(City of London, 1984, p. 16)

The plan sought to implement its conservation objective through the
designation of many Conservation Areas and through measures such
as plot ratio controls. The plan was also concerned to maintain a balance of uses and to retain certain activities that were considered to be
under threat such as housing, local services, small businesses, industry and warehousing. To ensure this balance certain restrictions on
offices were deemed necessary and justified in the following terms:
'an over-specialisation within certain economic activities and an
over-concentration on a certain type of worker could lead to economic and social problems in the future' (1984, p. 37). The plan could
be regarded as taking a fairly 'conventional' approach to the purpose
of planning, that is seeing the job as one of balancing all the varied
pressures on land use and seeking to ensure an end result that was
physically attractive.

However, the plan was published just as changes were being envisaged in the operations of the money markets. The advent of
deregulation, or the 'Big Bang' as it was called, had physical implications in the requirement for more spacious offices. The close
geographical links between activities in the City became less important. One indication of the new kind of demand was the huge Canary
Wharf proposal in Docklands, which represented a threat to the financial dominance of the City. Thus the careful balance of the 1984
plan was subjected to great pressure with demands for a relaxation of
the constraints on offices. The attack was led by Peter Palumbo, a
friend of Mrs Thatcher, who had had a scheme for Mansion House
Square turned down by the City. In commenting on the plan he said
that 'misguided conservation plans to preserve 70 per cent of the
centre of the City of London would reduce the world's leading commercial centre to little more than a museum of moderately
interesting buildings which a few tourists might occasionally glance
at' (*Planning*, 1985a). The draft plan attracted about 1,500 comments
of dissatisfaction, mainly regarding the restrictive office policies.
These were largely from the property sector (Debenham, Tewson &
Chinnocks, 1986). One of the main arguments expressed was that the
City's future depended on the continued supply of new high-quality
property and that there was a danger to the City's wealth if rates were
allowed to flow to surrounding boroughs. As a result of this pressure
a revised plan was produced in 1986 which sought to demonstrate the

City's positive attitude in supporting the new trends in financial development. The new plan relaxed considerably the conservation policies of the 1984 plan. However, this still didn't prevent the Associated Owners of City Properties and others from lodging objections to the new plan because the conservation policies were still considered too restrictive.

There is one particular objection to the 1984 draft plan that is of special interest. This is the submission from the Joseph Think Tank, the Centre for Policy Studies. This submission demonstrates the linkage between the opposition to the City plan and the broader ideological attack on planning. The submission took the form of a 44-page report which severely criticised the plan and recommended an almost total rewrite. However, the report also made many attacks and general comments on the role of planning. Early in their report the authors clearly set out their position on the role of planning and its relationship to the market:

> we believe in general – and above all in the City, the nation's financial market place – that development should respond to market forces, and that planning intervention should be limited to where it can be shown to be both necessary and effective to offset manifestly undesirable consequences of market forces.
>
> (Centre for Policy Studies, 1985, p. 2)

They complain that the 1984 draft plan unnecessarily interfered with market forces from a position of ignorance and also removed the rights of property owners.

One of the particular attacks on the plan was to suggest that it was totally out of date and lacked the knowledge and ability to keep abreast of current demands from financial interests. This clearly reflects ministerial attitudes. The inability of planners to foresee future demands was linked to their over-emphasis on conservation. The following quote demonstrates this line of thought:

> The proposals illustrate the shortcomings of the planning system. A range of detailed interventionist policies have been laboriously prepared over nearly a decade using outdated data, superficial analysis and outdated ideas by a body that does not understand the nature of what it is purporting to plan. The philosophy behind much of the plan unfortunately illustrates the national malaise of looking to the past not to the future, which is at the root of Britain's economic decline. The outcome demonstrates conclusively the failure of the cumbersome planning system when faced with a fast changing environment.
>
> (Centre for Policy Studies, 1985, p. 36)

The report also criticises the plan for intervening in the market to protect specific economic activities such as local services and small business. It is said that the plan produces no evidence to demonstrate that the market cannot provide these other activities and instead relies purely on assertion based upon 'current planning fashion'. In the manner of the academic writers reviewed in earlier chapters, the report claims that the result of this kind of intervention is to subsidise the economic activities that are getting this protection and to subject planning to pressure from vested interests. The report summarises its objection to the interventionist policies in the following way:

> The policies are neither justified nor are most of them feasible. The policies to favour fur and meat traders are based on purblind conservatism, lack of imagination, and a willingness to accede uncritically to sectional vested interests. The policies on industry, wholesaling, and housing are mere planning fashion. The City is an office location, pure and simple; policies that zone space for industrial, warehousing, and housing use are both irrelevant and undesirable.
>
> (Centre for Policy Studies, 1985, p. 22)

This examination of the City of London Plan has demonstrated a number of issues concerning the impact of New Right thinking on the role of planning. The need for the market to dominate the decision-making process and the facilitating role for planning is very clearly expressed in the Centre for Policy Studies objection (Centre for Policy Studies, 1985, p. 36). This market-oriented approach obviously reflects the concerns of Thatcherism. As a result a local plan which adopts a fairly traditional planning approach, attempting to balance all the competing demands, was rejected. It raises some interesting questions within the right-wing approach to planning. As shown at the beginning of this chapter, ministers constantly emphasise that they support the role of planning in maintaining the heritage and protecting the environment. The City case study shows how this can lead to conflicts when market forces come into opposition with the objective of conservation. The second interesting aspect concerns the issue of trying to ensure a mix of land uses in an area. The effect of the market in the City would appear to be to generate a purely office land use and this is encouraged by the Centre for Policy Studies Report. However, others would accept that this tendency of the market to produce dull homogenous areas can be economically counter-productive and hence require some kind of intervention. This was the view of Jacobs elaborated in the last chapter.

Unitary Development Plans and strategic guidance

The new initiative in relation to planning at the local level was the Unitary Development Plan. The Local Government Act 1985 abolished the GLC and the Metropolitan Counties of England and as a result a new system of planning, elaborated in Circular 30/85 (DoE, 1985d), was devised for these areas. The new system comprises a statement of strategic guidance prepared by the DoE and a two-part Unitary Development Plan, prepared by the local authority, which has to conform to the strategic guidance. The first part of the Unitary Development Plan is similar to a structure plan in content but is expected to be fairly brief, while the second part contains detailed policies similar to local plans and also a reasoned justification for both parts. The most important issues raised by this initiative, as far as the modification of the planning system is concerned, are central control and participation.

It has been suggested (Bristow, 1985a) that, for the government, the revision of the planning system implied by the Unitary Development Plan was a welcome spin-off from the abolition of the Metropolitan Authorities and an opportunity to downgrade strategic planning further. As Bristow puts it:

> Behind these moves one suspects the motive that most planning is unnecessary and too highly detailed, and that while development control work at the local level has its necessary uses, strategic thinking in current circumstances is generally unhelpful and can represent a challenge to central government authority.
>
> (Bristow, 1984a, p. 228)

Certainly the new system introduces a very different set of procedures for planning at the strategic level.

These procedures revolve around the preparation of the strategic guidance which will have statutory force and to which the Unitary Development Plan must conform. They introduce direct central government involvement in development decisions *before* plans are produced. The matters to be included in the guidance will be at the discretion of the Secretary of State and there will be ample opportunity for the government to impose its ideological perspective on the subsequent planning policies in the Unitary Development Plan. Each authority will have to show in the 'reasoned justification' of their plan how they have conformed to the guidance.

Participation in the preparation of the strategic guidance is far less than occurred under structure plans. Although views are to be sought on both the scope of the guidance and the draft guidance itself there is no obligation on the Secretary of State to take into account the views expressed. According to the GLC (1985), unless the Secretary of State decides to call in the Unitary Development Plan individuals, organisations will not be able to pursue their right to represent their views over the strategic aspects of the Unitary Development Plan. This is a downgrading of the participation opportunities previously available in structure plan regulations. In the South East considerable controversy has accompanied the preparation of strategic guidance but elsewhere reasonable agreement seems to have been reached between central and local government. However, such agreement does not alter the fact that central government has the power to impose its wishes if it so desires.

There will also be changes to the procedures in the local plan aspect of the Unitary Development Plan. There will be reductions in the flexibility available to authorities to devise different types of plans because under the Unitary Development Plan regulations authorities will have to produce a plan for their whole area and cannot produce a series of smaller ones. It has also been suggested (GLC, 1985) that there will be greater central government control and involvement in local plans through its reserve powers.

These views are illustrated in the conclusions that one London Borough draws from an examination of the new system:

> The role of strategic planning would be placed with the Secretary of State. His guidance to boroughs would be mandatory and not subject to examination in public.
>
> Each borough would still have to prepare its own 'structure' plan (as part of the Unitary Plan) and unless each plan was reviewed concurrently, there would be no coherent plan for the whole of London.
>
> The Secretary of State would be empowered to direct boroughs to modify their plans, even at a local level, in a manner to which they may be fundamentally opposed.
>
> (Lambeth London Borough, 1984)

This would suggest increasing conflict between central and local government in the operation of the Unitary Development Plan system and increasing uncertainty in the planning of our major cities.

This discussion of the Unitary Development Plan system at the local level has shown that there is concern over the purpose and objectives being pursued by some local plans. It is felt that insufficient attention is being given to market forces and that policies concerned with the protection of certain economic uses, protection of the environment and community interests need to be constrained in the interest of aiding the market and development. The Unitary Development Plan system in metropolitan areas will provide central government with greater control and reduce the opportunities for public involvement. As a result the operation of the market will have a greater potential role while political influence, whether from the community or particular threatened economic interests, will be reduced.

Lifting the Burden

On August 1st, 1984 Mrs Thatcher announced that an interdepartmental study would be set up to investigate the administrative and legal burdens on small business. The report *Burdens on Business* (Department of Trade and Industry, 1985) was published in March 1985 and led to the White Paper *Lifting the Burden* in July 1985 (DoE, 1985a). In May of that year the co-ordinator of the White Paper, Lord Young, spoke to the Association of British Chambers of Commerce on the theme of deregulation. Comments he made in this speech suggest that a fundamental review of planning lay behind these studies. He asked:

> Do we need the present system of town planning? The order regulating the use of buildings was introduced just after the war – and was based on industrial classifications of 1875.
>
> Is it any accident that the Minister for Sport and the Minister for Planning are one and the same? Our national pastime, it sometimes seems to me, is no longer cricket but public inquiries.
>
> (Reported in *The Times*, May 17th, 1985)

In the event the White Paper did contain a chapter devoted to town planning regulations and some major proposals for change. The proposals on Development Control aspects and Simplified Planning Zones will be considered later. This section will examine the implications for development plans.

The White Paper first states the government's intention to simplify the planning system and that, although much has been done, there is still scope for further progress. It also describes the purpose of the planning system as striking 'the right balance between the needs of development and the interests of conservation' (DoE, 1985a, p. 10). This is quite a restricted view of planning and as one London Borough has remarked, does not include community interest as an objective (Lambeth London Borough, 1985). Neither does it acknowledge, as the Conservative Party study groups did before 1979, that the public is an element in this balance. In the White Paper, development plans are seen as useful tools because they 'can assist developers and the business community by providing them with some indicators to guide them in taking their decisions' (DoE, 1985a, p. 14). However, this statement is not developed and instead the

Paper moves on to the shortcomings of the present system. This passage is quoted at length as it contains a number of important issues;

> Inevitably plans become out-of-date and tend to lag behind current needs and conditions. In particular, the twin priorities of generating jobs and providing sufficient land for housing have not been reflected fully or quickly enough in structure plans and the planning decisions of local authorities. The new circular issued by the Secretaries of State for the Environment and Wales (reproduced at Annex 2) accordingly makes it clear that development plans are one, but only one, of the material considerations that must be taken into account in dealing with planning applications. It is also important that development plans should concentrate on the essential elements and the key planning issues, be well related to current trends in the economy and the factors that influence market demand, and be capable of rapid revision to meet changing circumstances.
>
> (DoE, 1985a, p. 15)

The Paper continues by criticising the lengthy procedures of the development plan system and the excessive detail contained in structure plans and indicates that the government is reviewing changes to the system.

In the above quote we see the argument that has been constantly referred to by ministers at the RTPI Summer Schools, namely the inability of plans to keep up-to-date with contemporary needs and trends, carrying the implication that the best indicator of these current needs is the market. In the White Paper this argument is taken to the logical conclusion that the importance and ability of plans to determine decisions has to be reduced. Hence the plan becomes only one material consideration and the status of development plans is downgraded. Thus in determining an application for industry and business the plan is only one of the various considerations to take into account and cannot be seen to override other considerations. The principal 'material consideration' is expressed as 'the need to encourage employment and to provide the right conditions for economic growth' (Circular 14/85: DoE, 1985c, para. 5). The developer can therefore claim that his/her knowledge of the market should override the plan. Circular 14/85 states that at an appeal this kind of argument will be favourably considered. Another proposal is that plans have to be prepared more quickly and cover only essential elements, leading to less debate and participation.

Have development plans a future?

As mentioned above, the White Paper *Lifting the Burden* (DoE, 1985a) indicated that the government was looking at changes to the development plan system. This was confirmed in the Secretary of State's speech to the RTPI Summer School in September 1986 (Ridley, 1987). During the same month a consultative Green Paper, *The Future of Development Plans* (DoE, 1986b), was published. Before turning to this important paper reference will be made to two other reviews of the planning system that were published at this time, undertaken by the Royal Institute of Chartered Surveyors (RICS) and the British Property Federation (BPF). Both these organisations can be regarded as reflecting the views of developers and having some influence on the government.

The RICS report, *A Strategy for Planning* (RICS, 1986), starts from the premise that the planning system is still rooted in its post-war origins, oriented to the policy of restraint. Under current conditions the system 'must be better able to encourage and promote development'. The report identifies three fundamental problems with the current system:

(a) the influence of the public on the system;
(b) the lack of adequate, clear, defensible and robust policies;
(c) the growth of excessive 'political' influence at local level'.

(RICS, 1986, p. 3)

The writers of the report welcome the government's attempts to reinforce the view that there should be a presumption towards allowing development unless a good reason can be presented to oppose it. However, they claim that this is not consistent with allowing considerable consultation. They object to this consultation on two grounds, first, because it creates delay, and second, because it increases the 'range of considerations which need to be taken into account in the decision-making process' (1986, p. 3). They support the government's view that planning should stick to land-use issues and promote development. Consultation makes this difficult to pursue as it widens the debate and creates opposition to development. The report therefore proposes that the opportunities for consultation on plans need to be reduced. For example it suggests that major developments such as the third London Airport should not be debated at lengthy public inquiries but dealt with by Parliament. This is the line taken by the government on the Channel Tunnel. The report proposes that there should be no consultation at the strategic policy level and that it should be confined to local plans, 'since it is at this stage that the full practical implications of the policies should become evident and that sensible comments and objections can be expected' (RICS, 1986, p. 8). This consultation on the local plan should take place at the formal deposit stage. Once this has occurred the public should not be consulted again on individual applications.

The writers of the report then turn to the second problem they perceive in the system, namely its lack of clear policies. Rather than the Secretary of State trying to carry out policy formulation through the modification of plans that have been prepared by local authorities, they propose that the Secretary of State enter directly into the policy formulation stage through issuing regional policy guidance. They then look at the question of speeding up plan production, claiming that 'the development industry needs, and will continue to need, a much more rapid response in terms of policy guidance' (RICS, 1986, p. 7). They say that policies should be monitored annually and plans updated every two years and that to do this procedures need to be simplified. The involvement of the Secretary of State in guidance and the reduction of participation should contribute to this simplification.

The report then turns to the problem of political influence. It says that the trend towards incorporating social goals in planning has had the detrimental effect of allowing elected representatives to impose their political views, often against the advice of professional planning officers. Although they admit that 'it is within both the letter and the intention of the law that elected members should have this power in the interests of democracy, decisions that run counter to accepted good practice often result in successful but costly appeals' (1986, p. 9). The authors of the report considered removing the responsibility for development control from councils to an independent body to provide 'a greater measure of impartiality, predictability and rationality' (1986, p. 9). However, instead they recommend greater delegation to planning officers.

There are several themes that run through these suggestions and which express quite clearly the attitudes behind the proposed changes. To the extent that some of these changes also appear in the government proposals, with rather less explicit reasoning, the RICS report is useful in suggesting some of the attitudes behind the government's own proposals. One of the themes is allowing developers greater scope in carrying out their schemes. This is reflected in the greater status to be given to market trends, the reduction in participation, and the attempt to remove political interference. There is also a clear downgrading of the importance of democracy at the local level with greater reliance on Parliament and central direction and control. The attempt to speed up the system also has the effect of shifting this balance from wide participation towards central direction and market criteria.

The British Property Federation Report, *The Planning System – A Fresh Approach* (BPF, 1986), was published in May 1986. The view of the BPF is that Local Planning Authorities (LPAs) have too much power and that this disadvantages developers. The aim of their proposals is to shift the balance back to developers through the importance given to economic and commercial factors. In their view economic initiative and regeneration are of prime importance and can only be given due regard through the market process. Thus the planning system needs to be pruned back and curtailed. 'It should not be part of the role of a Local Planning Authority to interfere with the normal ebb and flow of land use demand and supply' (BPF, 1986, p. 25). The complex bureaucracy and inflexibility of an LPA cannot match the sensitivity of the market with the result that 'the attitudes and policies of LPAs are not in tune with the needs of present-day society' (1986, p. 6). In order to remedy these faults planning decisions should be speeded up, the economic costs of delay and conditions taken into account and plans renewed frequently to keep up with market changes.

A second major area of concern to the BPF is the uncertainty generated by the existing planning system in Britain and they examine alternative systems in West Germany and the USA. They see considerable benefits accruing from the zoning approach in these countries as it allows developers to know beforehand with greater certainty what is acceptable and removes the vagaries caused by the discretion inherent in the British system: 'developers know that, provided they comply with the zoning ordinance and the building code ordinance, they can go ahead with their plans confident that approval will be forthcoming' (BPF, 1986, p. 8). In a proposal that echoes the earlier ideas of the Adam Smith Institute, the BPF report proposes a zoning system made up of three kinds of zones. The first would be called Stability Zones and include areas, such as housing or agriculture, where the existing economic use was healthy and prospering. In these zones the existing planning system, as modified by their other suggestions, would apply. Second, there would be Activity Zones which would include all commercial and industrial areas and areas where the existing use is undergoing change. Planning controls would be freed up and operated through the mechanism of a 'specification'. These zones would be similar to the government's Simplified Planning Zones but would be more widely applied and more permanent. However, in contrast to the Simplified Planning Zone planning permission would still be needed under the Activity Zone. The third category, called Controlled Activity Zones, would be rarely used but would cover areas requiring greater control such as centres of historic towns. In these zones a more detailed specification would be formulated.

It is proposed that the 'specification' becomes the primary tool for planning control. This specification, part of the plan, would be limited to matters of land use, appearance, impact on adjoining land, and aspects of the development such as car-parking or landscaping. The report considers that design should be left to the architects who are acting for the developer. In general the report seeks to impose much greater constraints on the scope of planning, especially to remove 'social engineering' or political factors.

A final theme that can be drawn from the report is the concern, also voiced by the RICS, about the effects of the democratic process. The BPF view is that although democracy is necessary it should not be allowed to stifle change. Democracy therefore needs to be kept in balance with other factors, namely planning expertise, central government and commerce. It therefore makes suggestions to strengthen the influence of these other factors on planning committees. It proposes that a sub-committee of the planning committee be required called the Business Advisory Committee and that the views of this sub-committee should be taken into account in any appeal. The report also believes that Chief Planning Officers are subject to too much political pressure. The professional expertise of the Chief officer and his/her responsiveness to central government needs to be given more weight. The mechanism for doing this is to make 'both the appointment and the dismissal of Chief Planning Officers ... subject to the approval of the Secretary of State' (BPF, 1986, p. 12).

It is now necessary to examine whether the government's proposals bear any resemblance to those of the RICS and BPF and hence whether the underlying attitudes of these two reports can also be attributed to the government.

The Green Paper, *The Future of Development Plans* (DoE, 1986b), reinforces the previous statements by the government that planning has the dual purpose of helping to promote economic generation while also conserving the best heritage. It says:

> The planning system has to cater for a diverse and market-related pattern of economic activity. It has to facilitate economic development and employment opportunities. It has to respond to rapidly changing technology and to major changes in retailing, in manufacturing and in the use of leisure. It has to ensure that adequate provision is made for land for housing, making full use of derelict and vacant land in urban areas. The need is for a system which is flexible and responsive in providing for these changes but which maintains its protection of those areas whose continued

conservation is important to the future quality of life in Britain.
(DoE, 1986b, p. 3)

Once again there is no mention here of the community interest or broader social considerations. The emphasis is very much on responding to the dynamic market with the proviso that some areas may require protection. The need for swift responses leads to the government concern to maintain flexibility in the system while also improving its speed of operation. For this reason the replacement of the existing system with a less flexible zoning system is rejected in the Green Paper in favour of modifications.

The Paper identifies three basic problems with the existing system: the scope of the policies, the relationship between structure and local plans and the complexity of the procedures. The Paper continues the theme established in the Secretary of State's modifications to structure plans, that plans are too broad in scope and over-detailed. Structure Plans include 'policies that have nothing to do with land-use planning or improving the physical environment' (DoE, 1986b, p. 7). The Green Paper gives examples of the kind of policies that should not be included in structure plans and, perhaps, not in development plans at all. These examples are:

> building design standards, storage of cycles, the costs of waste collection, the development of co-operatives, racial or sexual disadvantage, standards of highway maintenance, parking charges, the location of picnic sites and so-called 'nuclear free zones'.
> (DoE, 1986b, p. 8)

The influence of the proposed alterations to the Greater London Development Plan (GLDP) on government thinking can be detected here. Sir George Young is quoted as saying, 'the contents of the proposed [GLDP] alterations emphasise the point that they include such matters as policy for women and community areas which, while important, can hardly be regarded as relevant to a land-use structure plan' (May 1984 quoted in GLC, 1985, p. 7). Local plans are also criticised for containing irrelevant or over-detailed policies. Other problems with local plans are the delays created by their relationship with structure plans and the increasing use of informal plans and supplementary guidance which creates confusion.

Although some modifications were made, the general philosophy and approach of the Green Paper was carried through into the White Paper published in January 1989 and the Planning Policy Guidance Note on Local Plans published a few months earlier. The first proposal in the White Paper is that structure plans should be abolished. The Secretary of State will prepare regional planning guidelines which local authorities will have to conform to in their planning work. This was also proposed by the RICS report and can be regarded as extending to the rest of the country the strategic guidance aspect of Unitary Development Plans. Counties will then prepare policy statements on matters that cannot be dealt with at the local level. However, this will be restricted to key topics. The Secretary of State will specify what these matters are and the counties will have to prepare their statements so that they conform to the Secretary of State's regional guidance. Local plans will continue in much the same form as at present although they will have to conform to regional guidance and county statements and avoid unnecessary detail and irrelevant policies. All informal plans and supplementary guidance will cease to be prepared and will not be given any weight in appeals.

The White Paper then describes the new procedures required. As far as county statements are concerned the statutory procedures would be confined to the regulation of their scope, the requirement to publish a draft for comment and to hold an examination in public and the process of formal adoption. The Secretary of State can require modifications to, or call-in, the proposed county statement. The new local plan procedures are based upon existing ones but streamlined to shorten the time taken to complete. A reduction in participation is proposed. It is suggested that authorities should be encouraged to involve the public before a draft plan is prepared perhaps through consultation on issue papers. There would be no requirement to participate with the public over the draft plan but the six-week period for representations and objections after the deposit of the plan would be retained. The local planning authority would no longer have to prepare a report on the steps they had taken to involve the public and submit this to the Secretary of State. As with county statements the Secretary of State will have the power to require modifications to, or call-in, the plan.

Some of these proposals reflect those suggested by the RICS, for example greater central control and involvement of the Secretary of State in policy formulation. The Secretary of State also has considerable reserve powers of modification and call-in. There is also the concern to reduce participation, although the details differ from the RICS Report. The government's proposals on local plan participation have been criticised for concentrating participation on a stage that experience has shown produces little response. Thus the proposals are likely to result in a major reduction in involvement as this would be concentrated on the formal stage after the plan had been produced. It must be assumed therefore that the government desires less participation although the reasons for this are not made explicit.

In the White Paper there is no direct reference to political interference in contrast to the RICS and BPF reports, although the continual stress on the need to confine plans to land-use issues could have the effect of bringing decision-making closer to the 'professional issues' desired by both these bodies.

Conclusions

The modifications through the period have left development plans considerably weakened. Structure plans were undermined by the Local Government, Planning and Land Act 1980 and replaced by the Secretary of State's strategic guidance in London and the Metropolitan Counties. In the White Paper *The Future of Development Plans* (DoE, 1986b) it is proposed to extend this system and abolish structure plans. Meanwhile all plans were downgraded in the *Lifting the Burden* White Paper (DoE, 1985a) and Circular 14/85 (DoE, 1985c) to become 'only one material consideration'.

A number of themes can be identified as running throughout these changes and these will be examined in relation to the framework set out in Chapter Four. The first aspect of this framework refers to the **principles** of decision-making. As the RTPI Summer School speeches illustrate, the intention of government is to retain the bones of the planning system but to give it a new shape and purpose. This purpose is one which has as its primary aim that of aiding the market. The planning system must keep up with the current trends in that market and foster and nurture them. The modification to the purpose of planning amounts to giving the developer greater freedom. This market-led approach is illustrated in nearly all the initiatives examined in this chapter and the direct involvement of central government in strategic guidance can ensure this emphasis throughout the system.

The modifications to structure plans early in the administration's life concentrated on removing the ability to incorporate social factors and this has also been reflected in comments on local plans and in the White Papers *Lifting the Burden* and *The Future of Development Plans*. This reduction in scope is reinforced by the lowering of importance of development plans as a 'material consideration' and Circular 2/87 (DoE, 1987b) states that costs will be awarded against a local authority if it rigidly adheres to its development plan ignoring other 'material considerations'. The Circular implies (para. 8) that both the merits of particular applications and changed circumstances can override the approved plan.

The two 'acceptable' criteria, that is economic promotion and amenity preservation, do however come into conflict at times. This

143

was demonstrated in the objections to the City Plan where it was considered that the amenity preservation criterion had been taken too far. There is therefore still potential within the new restricted scope of planning for uncertainty and disagreement to arise.

The second dimension of the framework was that of the **procedures** adopted. Here two aspects can be identified: the increase in centralisation and the reduction in participation. Central control has the effect of reducing the ability of local authorities to introduce their own criteria and allows central government to ensure that market criteria dominate. Increased central control is evident in the early structure plan modifications, the facility of call-in powers, the device of strategic guidance, the use made of Circulars and the importance given to them at appeals.

The second procedural aspect, which in part flows from the importance given to the market, is the reduction in participation. As the RICS has mentioned, commercial criteria and participation are not compatible as increased involvement and consultation is bound to broaden the range of criteria covered in the discussion. Again, the reduction of opportunities to participate is a thread running through the period associated with the desire to speed up and streamline the system. These reduced opportunities can be seen in the Local Government, Planning and Land Act 1980, the Unitary Development Plans, and the White Paper *The Future of Development Plans* (DoE, 1986b).

The RICS and BPF proposals take some of the themes a stage further. Most of their ideas reflect government thinking, such as the extension of central control through the wider use of the Unitary Development Plan system or the reduction in the scope of plans. However, they do express the anti-democratic theme rather more explicitly. The RICS considered removing development control from the political arena and recommends greater delegation to the chief planning officer. The BPF wants to establish Business Advisory Committees with power exerted through the appeal system and central government appointment of chief planning officers. All these ideas are intended to reduce the influence of local democracy.

The two reports also raise questions about attitudes to administrative discretion and a preconceived general framework such as Hayek's 'rule of law'. It can be said that many of the initiatives covered in this chapter would reduce administrative discretion both through the reduction in scope of planning and through the centralisation process. The BPF places considerable emphasis on the need to remove uncertainty and for this reason prefers the zoning system of the USA and West Germany and suggests its own zoning system for this country. This last suggestion is clearly a move to a preconceived generalised framework. It will be necessary to return to these issues once the Urban Development Corporations and Simplified Planning Zones have been explored in later chapters.

So development plans have not been completely abolished even though structure plans are under threat. After a period of uncertainty there is a firm commitment to local plans for all areas. However, this formal retention masks a significant shift in power and control towards central government and market forces. The freedom and scope of the plans are considerably diminished. County statements are to be limited to key issues, informal plans are not to be used and the formal ones must stick to narrowly defined land-use aspects. Plans can be challenged by claiming the greater relevance of other 'material considerations' and in some areas central government has the ability to impose its values on the whole system through its strategic guidance.

Chapter seven

Modifications to the planning system:
2 Development control

This chapter continues the exploration of the modifications to planning legislation by looking at the development control aspect. Particular attention is given to Circular 22/80 (DoE, 1980b), which has been referred to as the most significant statement of the government's approach and which also provides the context for many of the later changes. The discussion of the deregulation of planning control then continues through government statements on 'planning gain', planning conditions, relaxing controls on industry and small business, and the review of the General Development Order and the Use Classes Order. The cumulative effects of these various modifications are discussed and related to the themes of Thatcherism identified in earlier chapters.

This book focuses on the planning legislation itself rather than the organisational processes for implementing it. However, one cannot discuss development control over the last decade without mentioning the issue of 'delay'. This has had a pervasive influence throughout the period and has affected the manner in which both existing and new legislation has been put into practice. One of the first concerns of the new government in 1979 was to speed up the development control system. Heseltine made his well-known reference to jobs being locked away in planners' filing cabinets. The view that was propagated stated that delays created 'extra costs, wasted capital, delayed production, reduced employment, income, and profitability' (DoE, 1988a, para. 6). Planners were urged to mend their ways.

The government said that planning applications should normally be decided within eight weeks and set time targets for handling applications and appeals. They also published 'league tables' of local authority performance in meeting these targets which led, in turn, to many local authorities producing similar statistics for individual officers. However, the approach is very simplistic and does not account for the variety and complexity of applications (see for example,

Thompson, 1987). It has the overall effect of putting pressure on officers to make quick decisions and cut corners. This can mean a reduction in the range of considerations covered and a streamlining of any participation and involvement. Such implications are also evident from a detailed examination of the various Circulars issued since 1979 to guide development control.

The centrepiece: Circular 22/80

The first significant modification to the development control aspect of the planning system came with Circular 22/80 *Development Control – Policy and Practice* (DoE, 1980b) published in November 1980. This had been preceded by several months of controversial debate over the draft Circular. In the event the content of the draft was retained although the tone was moderated. Circular 22/80 is an important statement of the wish of the government, at an early stage in its period of office, to make changes to the planning system. In 1984 it was said that 'no Circular emanating from Marsham Street in recent years has evidenced the changes in principle and emphasis that have occurred in central government thinking about the role of the planning system as much as Department of Environment Circular 22/80' (Brand and Williams, 1984, p. 610).

It has been noted (for example Rydin, 1986) how influential the housebuilding lobby was in the drafting of this Circular, particularly through the personage of Tom Baron, Honorary Secretary of the Volume Builders Study Group, who was seconded to the DoE at the time. It has been shown above how modifications to structure plans by the Secretary of State often led to an increase in the allocation of land for housing. This was reinforced by Circular 9/80 (DoE, 1980a), which required local authorities to utilise land availability studies involving private housebuilding interests. Rydin has suggested that the net effect of this involvement of the private housebuilding lobby in structure plan changes, land availability studies and Circulars 9/80 and 22/80 has not been simply a loosening up of the planning system and the release of more land for development but rather a shift in the *control* of decision-making.

The importance of the 'Circular' as an instrument of central government policy should perhaps be mentioned at this point. In theory Circulars are not statutory and are merely advisory documents and therefore local authorities can apply them with discretion. However, as stated in the final sentence of Circular 22/80, when 'appeals are made the Secretary of State will be very much guided by the policy advice set out in this Circular'. The knowledge that an applicant can go to appeal and utilise the Circular in support of his/her case will very

147

much circumscribe local authorities. In an editorial in the *Estates Gazette* (December 6th, 1980) the inspectorate is urged to ensure that the policies in Circular 22/80 are given force at appeals as many local authorities may not be inclined to implement the Circular. Furthermore it has been shown in two cases that went to the High Court that if the Inspector or Secretary of State does not fully take into account the Circular in deciding the appeal then the court is likely to override the decision (Rydin, 1986, p. 79; Purdue, 1982, pp. 579–81). This would seem to give the policy in a Circular considerable legal status and hence power.

The aims of Circular 22/80 are to speed up the planning system and to make it more responsive to development. In particular, planning should 'create the right conditions to enable the house building industry to meet the public's need for housing' (para. 3) and 'bear in mind the vital role of small-scale enterprises in promoting future economic growth' (para. 11). Local planning authorities are reminded always to grant planning permission unless there are clear-cut reasons for refusal. There are a number of aspects of the recommendations that are important in exploring the modification of the planning system; they are the reduction in the criteria to be used in making a decision, the importance of market factors, comments on zoning, and the implications for participation.

The Circular develops the view, expressed by the Secretary of State at the 1979 RTPI Summer School (Heseltine, 1979), that planners should withdraw from aesthetic control, 'and should not impose their tastes on developers simply because they believe them to be superior' (DoE, 1980b, para. 19). The Circular suggests that control of external appearance can be important in environmentally sensitive areas such as conservation areas or National Parks but that elsewhere design details should be left to the developer and his/her architect. The fact that the developer will have to sell the development on the market will be the control over acceptable design and will also allow greater innovation, not to mention reduction of delay. In the years following the Circular the Secretary of State was concerned that local authorities were not taking the relaxation of design control seriously. In 1985 another Circular was issued (31/85: DoE, 1985e), which expressed the view that too many applications were subject to delay because local authorities were still involving themselves in detailed design matters and that a large number of appeals involved design aspects. The new Circular quoted sections of Circular 22/80 and reiterated their importance.

The draft to Circular 22/80 also specifically advised planning authorities not to attempt to influence the layout and density of housing schemes (*Planning*, 1980c). Brand and Williams (1984) report how the Circular has been used at appeal to reverse refusals that were based on the size of rooms in a residential development and the mix of housing type in a development. In this last case the inspector drew attention to the Circular in saying that 'local planning authorities should regulate the mix of housing types only where there were specific planning reasons for such control and that in so doing they should take particular account of marketing considerations' (Brand and Williams, 1984, p. 611).

This last point of taking into account marketing considerations in making a decision underlies both Circular 22/80 and 9/80 (DoE, 1980b, 1980a). It has already been noted how the government considers that planning practice should help housebuilders meet public demand. Circular 22/80 also requests that applicants should not be asked to 'adopt designs which are unpopular with their customers or clients' (DoE, 1980b, para. 19). As Rydin (1986) suggests, the importance of market demand in reaching a decision in effect places power and the control of the decision in the hands of the housebuilder who can claim to be the judge of the marketability of the development.

Another aspect of Circular 22/80 worthy of attention is the stance taken on the control of non-conforming uses. Relaxation is advocated. This is taken up in relation to small businesses which are considered to be adversely affected by planners' rigid zoning of industrial and residential areas. The Circular states that there are many businesses that can take place in residential and rural areas without causing disturbance and therefore concludes that 'the fact that an activity is a non-conforming use is not a sufficient reason in itself for refusing planning permission or taking enforcement action'. Planning authorities are requested to assume that when small-scale industrial or commercial activities are proposed in residential or rural areas permission will be granted unless there are specific objections. The kinds of objections mentioned are intrusion into open countryside, noise, smell, safety, health or excessive traffic generation. It has been suggested (for example, *Estates Gazette*, December 6th, 1980) that this relaxation of control does not go far enough and that these specific objections still give planners too much scope for refusal.

The final aspect to note is the implied attitude to democracy and participation taken in Circular 22/80. It has been seen that the Circular wishes to relax the control over design by the local authority. In putting forward this view the Circular adopts Heseltine's opinion that design standards and democracy are not necessarily compatible. He says,

democracy as a system of government I will defend against all comers but as an arbiter of taste or as a judge of aesthetic or artistic standards it falls far short of a far less controlled system of individual, corporate or institutional patronage and initiative.

(DoE, 1980b, para. 18)

Underlying these remarks is a strong body of opinion which believes that politicians are unduly influenced by public pressure, particularly by those only marginally affected by the decision (*Planning*, 1980b). It will be noticed that this is an attitude which is very similar to that of Siegan and others outlined in Chapter Five.

The Welsh Local Ombudsman, in his annual report for the year ending March 1981, is critical of the participation aspects of Circular 22/80 because it reduces the opportunities for individuals, whose lives and homes may be affected by a proposal, to express their views. Although accepting that protection of neighbours' interests may not be the primary function of a planning authority, the Ombudsman sees dangers in the speeding up of the system and the reduced control over the appearance of buildings (reported in the *Journal of Planning and Environmental Law*, November 1981, p. 779).

There have been a number of reports that have tried to analyse the impact of Circular 22/80 on appeal decisions. Drawing on a number of these, thirty-three appeal decisions can be examined in the three years after the Circular was issued (evidence is drawn from *Planning*, 1981a; 1981b; 1981d; 1982; 1983b; Brand and Williams, 1984). A narrowing of the range of criteria used to refuse permission can certainly be detected and well over half of the reported cases show how Circular 22/80 was successfully used to win an appeal. However, much uncertainty remains. It seems that the line is fairly clear in excluding detailed design matters. The real area of difficulty appears to be deciding between, on the one hand, business and housing needs and, on the other, protection of amenity. Cases of businesses in rural areas have gone either way at appeal and there are a number of examples where housing demand has been rejected because of environmental objections. One of the most contentious areas seems to be non-conforming uses in residential areas and here there is much contradiction and lack of consistency in appeal decisions.

The general picture is one of increasing weight being given to the needs of small businesses and housing land with environmental aspects often being overridden. However, sometimes amenity arguments have been successful, for example in conservation or new housing areas. A further three cases reviewed in 1985 showed that the Secretary of State often ignores the studies on availability of housing land and allows an appeal where market forces show that the location is attractive to buyers or there is a demand for the particular

kind of housing proposed. This demonstrates the overriding importance given to the criteria of market demand (*Planning*, 1985b).

There are a number of conclusions that can be drawn from this important Circular. Market criteria have been given greater status and
planning control over design matters has been significantly reduced.
However, although the needs of housing development and business
are paramount, difficulties arise where these needs conflict with environmental issues. A two-tier system is created as areas of sensitive
environments are excluded from the relaxation of controls. The reduced controls, the desire to speed up the process, and the increased
involvement of central government through appeals all reduce participatory opportunities.

Further restrictions: the use of planning conditions

In the wake of Circular 22/80 with its removal of planning control
over detailed design, the House Builders Federation began, in 1981,
to press for a review of the 1968 Circular on planning conditions.
Mervyn Dobson, their land and planning officer, described how the
Federation was unhappy about the way in which the planning control
system operated and in particular the way in which many authorities
over-used conditions and contradicted the philosophy of Circular
22/80. He said, 'there is no doubt that over the course of the last decade the level of control exercised through planning conditions has
become more wide ranging and detailed than ever before' (Dobson,
1981). Amongst his complaints were that conditions covered matters
that were not relevant to planning, that they were too complex, too
numerous or too vague. Eventually most of these concerns were
taken up in a draft Circular in November 1983 and a final version,
Circular 1/85 – *The Use of Conditions in Planning Permissions* – was
released in January 1985 (DoE, 1985b).

This Circular stresses that conditions should be used in a 'reasonable way' and seeks to set out tests for this. Broad conditions such as
'to secure the proper planning of the area' or 'to protect amenity' are
considered too vague (DoE, 1985b, para. 8). All conditions have to
be precise, relevant to planning and to the particular development
proposed. Conditions should not be used unnecessarily; 'a condition
ought not to be imposed unless there is a definite need for it and
planning permission would have to be refused if that condition were
not to be imposed' (para. 12). In other words a condition cannot be
used to improve a proposal from a merely passable scheme to a good
one. Conditions should not be too onerous, for example making it
difficult to run a proposed business or difficult to sell a property. Circular 1/85 also expresses a preference for the use of conditions rather

than Section 52 agreements so that the developer has the opportunity to appeal.

Planning gain: a loophole?

The establishment of planning agreements, using Section 52 of the Town and Country Planning Act 1971, between local planning authorities and developers, in which developers agree to provide certain facilities suggested by the authority, has become a common feature of the development control system. The local authority obtains 'planning gain' from the development and the developers obtain planning permission with, it is hoped, certainty and speed. There has been quite widespread acceptance of the value of this process, particularly for the flexibility it provides in the development control system. However, it has also raised many controversial issues. Even amongst supporters of planning gain, concern has been expressed over the secret and undemocratic nature of the process and the potential for abuse whereby permission could be granted to an inherently 'bad' development if sufficient planning gain was offered. It is not intended to review fully the debates on planning gain but simply to identify changes since 1979 and their implications (for a fuller discussion of the arguments about planning gain, see for example, Simpson, 1987; Jowell and Grant, 1983).

Notwithstanding the doubts there has been increasing interest in the use of negotiation and agreements. However, the ability to employ this approach is restricted to areas where there is pressure for development and hence where concessions can be extracted from potential developers. One of the frequently expressed advantages of the planning agreement approach is that it allows the authority to extract some of the development value created by planning permission and use it for community benefit. Other means of doing this, such as discussed in the Uthwatt Report or in the Community Land Act, are not now available. It is therefore one of the few opportunities to carry out an element of 'positive' planning. It can be viewed as a means by which social and economic factors can be introduced into the planning decision. As the scope of planning is reduced and the definition of 'material considerations' narrowed then the planning agreement is seen by many authorities as a welcome means of bypassing these restrictions. The 'normal' development control processes are subject to the constrictions imposed by Circulars, appeal decisions and legal interpretation, all of which have pointed since 1979 towards a reduction in the range of criteria that can be employed in development control decisions. Meanwhile planning agreements have not been subject to the same degree of supervision (see Simpson, 1987).

Against this background the government asked the Property Advisory Group to look at the issue of planning gain and their report was published in October 1981. This report was very antagonistic towards the process and concluded that, with two exceptions, the practice of 'bargaining for planning gain should be firmly discouraged' otherwise 'the entire system of development control becomes subtly distorted and may fall into disrepute'. The two exceptions were first, when a requirement could not legitimately be included in a condition, for example infrastructure provision off the site and, second, in a mixed development where some aspects in isolation would be regarded as a reason for refusal but could be offset by other elements which brought positive planning objectives.

The government took a long time to respond to this report and it was not until March 1983 that they produced their draft Circular, which was later issued in August, with minor changes, as Circular 22/83 (DoE, 1983a). This was far more accepting of planning agreements than the Property Advisory Group Report and said that such agreements 'may well assist towards securing the best use of land and a properly planned environment'. However, the Circular did try and set out guidelines for the use of agreements and these would be used at appeal. Circular 22/83 also threatened to make authorities pay legal costs if they made unreasonable demands on developers. Various tests of 'reasonableness' are set out. These tests are intended to 'provide guidance to local authorities about the proper limits of their statutory development control powers' (para. 13). Obligations on developers should be restricted to:

1 actions necessary for the development to occur, for example, provision of facilities;
2 financial payments towards such facilities;
3 facilities clearly related to the proposed development or its future use, for example car parking or open space;
4 arrangements aimed at achieving a balance of uses in a mixed development.

In addition two further tests are proposed. First, that the requirements should be reasonably related in scale to the proposed development and, second, that any financial payment should be reasonably imposed on the developer rather than through national or local taxation. Examples are also given of instances where the use of agreement is inappropriate, for example maintenance costs or the demolition of buildings unrelated to the proposal.

Concern has been expressed over the Secretary of State getting involved in monitoring planning gain and providing guidelines. This has been seen as detracting from a local authority's autonomy and ability to be flexible. Underlying this comment is the fear that the restrictions that apply to the 'normal' planning process would be extended to planning agreements, for example the tests in the Circular were seen by the Association of Metropolitan Authorities as too restrictive (Johnston, 1983b). Brownill (1989) has described several planning agreements in London's Dockland. She identifies a number of problems with these: their secrecy, they often involve bargaining away policy objectives and the 'gains' are often false, for example they may include facilities which would have been provided anyway. The effect of these agreements is to reduce public involvement while also reducing public opposition to the schemes, to provide the developers with a reputation for helping the community and to prevent any strategic approach. The London Borough of Ealing has tried to overcome some of the problems with planning gain by formulating a preconceived set of principles to apply to all large developments (Knibbs, 1989).

Some developers are worried that planning gain gives local authorities too much scope to introduce social criteria as they increasingly use planning gain to circumvent the restrictions on the use of conditions. However, so far the government has felt obliged to retain the principle of planning gain although restricting its operation via guidelines. The controversial nature of the issues raised by planning gain is likely to guarantee continued debate. In July 1989 the DoE issued new draft guidance and proposed amendments to section 52 of the Town and Country Planning Act 1971. This new guidance confirmed the code of practice over acceptable deals. It also introduced a new proposal to allow a developer the option of giving a unilateral undertaking when making a planning application. It would not be necessary for a local authority to agree the terms of this undertaking and the authority and the Secretary of State would have to take it into account in deciding on an application or appeal. The initiative would therefore shift in the direction of the developer.

The enterprise culture: relaxing controls on industry and small business

During 1984 and 1985 the government took further steps to try and reduce the control over industry, particularly small business. There was considerable inter-departmental examination of the issue and a report, *Burdens on Business* (Department of Trade and Industry, 1985), was published in March 1985. As already mentioned, this report provided the basis for the White Paper *Lifting the Burden* published later in the year (DoE, 1985a). Two Circulars were also issued: Circular

16/84 *Industrial Development* (DoE, 1984a) and Circular 14/85 *Development and Employment* (DoE, 1985c). The Circulars are introduced as an expansion of the advice in Circular 22/80 (DoE, 1980b). They state that priority should be given to industrial development and that such development should always be allowed unless it 'would cause demonstrable harm to interests of acknowledged importance' (Circular 14/85: DoE, 1985c, para. 3). These interests are stated as protection of heritage, environmental improvement, Green Belt and good agricultural land. The need to reduce delay, avoid complex and unnecessary conditions and accept light industry in residential areas are all again stressed.

A review (Holt, 1986) of the effect of Circular 14/85, as expressed in appeal cases carried out three months after it came into force, showed that it had a tendency to tip the balance slightly towards development, particularly for small business, and was a further extension of the relaxation contained in Circular 22/80. However, where highway, amenity and heritage protection were relevant factors, control was still strong. One of the cases quoted by Holt shows that development plans can be regarded as out of date even if they were prepared within a few years of the appeal.

More recently Circular 2/86 has mentioned the need to be more flexible in dealing with speculative building for small businesses; 'it is for market forces to determine whether there is a demand for such premises' (DoE, 1986c, para. 7). Circular 2/86 reiterates the point made in *Lifting the Burden* that planning should not try and interfere in the operations of the market:

> Planning control is not intended to enable local planning authorities to intervene in the normal operation of market forces, and local planning authorities should not refuse permission for new development by small businesses on the grounds that the proposed use would adversely affect the trade of established businesses.
>
> (DoE, 1986c, para. 11)

Circular 2/86 also stresses that part of a dwelling house can be used for a business as 'long as it does not change the overall character of its use as a residence' (para. 4).

The White Paper *Lifting the Burden* (DoE, 1985a) also raised the issue of the need for flexibility in dealing with high-tech firms. According to the White Paper these are firms 'where manufacturing, offices, research and development, warehousing and other activities may be carried on in a single building and where the mix of uses and space utilisation may need to be constantly changed and adapted to the needs of the business' (DoE, 1985a, p. 12). It is suggested that the review of the Use Classes Order will need to take this into account. Another response to this issue came in a further White Paper, *Building Businesses – Not Barriers* (Department of Employment, 1986), published in May 1986. This Paper put forward the idea of 'flexible planning permissions'. This proposal allows permission to be given for more than one use. Subsequently if changes take place within these permitted uses then no further planning permission would be required.

Further deregulation: changes to the General Development Order and the Use Classes Order

The General Development Order (GDO) specifies which types of development can take place without planning permission and thus, as the White Paper *Lifting the Burden* points out, can be used 'as a useful means of deregulation within the planning system'. In 1981 the government made changes to the GDO. In essence these changes allowed householders, except those in terrace housing, to extend their houses to a greater extent than was previously allowed without the need for permission. It allowed larger extensions for industrial premises and change of use from light industrial use to warehouse use, and vice versa, for small units. These relaxations do not apply in conservation areas, National Parks and Areas of Outstanding Natural Beauty. The changes have not been seen as controversial and in general terms follow the 1977 proposals of the Labour government. (For details and comment see Swann, 1980; *Planning*, 1981c, 1980a, and 1980d.) It could be said that the increasing complexity introduced has counterbalanced the intended effects of deregulation and also led to tightened controls in some areas such as terrace housing in inner cites.

As part of its deregulation package in the White Paper *Lifting the Burden* the government proposed further relaxations of the GDO. These were particularly aimed at giving greater freedom to business and increased the size of extensions to industrial buildings and warehouses that would be allowed without permission. In 1988 a new GDO was published accompanied by Circular 22/88 (DoE, 1988e) giving guidance on its application. It states that this new GDO seeks to simplify the planning system. It provides greater controls on certain buildings in National Parks, Areas of Outstanding Natural Beauty and conservation areas while removing certain other controls on commercial and industrial changes of use. This illustrates the general trend of relaxing controls and allowing greater market freedom while also strengthening conservation in selected areas.

Lifting the Burden also indicated that a review of the Use Classes

Order (UCO) was under way. The UCO had not been reviewed since it came into operation in 1948 and for many years there has been pressure for its revision because of its dated categories and its inadequacy in implementing planning policies (see for example, West, 1983; Kirby and Holt, 1986). However, the government's review was set within the context of a reduction in planning control and the Property Advisory Group were given the job of writing a report on the matter.

This report, which was published by the DoE in December 1985, viewed the UCO as a means of liberalising 'change of use' from planning control alongside other liberalising approaches such as Simplified Planning Zones and 'flexible planning permissions'. Its conclusions contained a number of radical proposals which if implemented would significantly diminish the power of planning to control land use (Holt, 1985). The most important of the proposals in this respect were:

1 To combine previous categories so that planning control is removed from change of use from shops to other uses such as banks, building societies and estate agents. Most of the Planning Advisory Group also wanted to extend this to include professional services such as lawyers, doctors and surveyors.
2 To amalgamate office and light industry into one category, called general business use, thus allowing change of use between them to occur without the need for planning permission.
3 To free up controls on residential use so as to allow small business (up to five employees) to occur in the home and give owners the right to subdivide their property without requiring permission.

There was a considerable reaction to these proposals (see RTPI, 1986; *Planning*, 1986a; Holt, 1985; Home, 1987). The Tory-controlled Borough of Westminster was among the critics of the proposals and it might have been pressure of this kind that led the government to water down the proposals in their report which was published for consultation in June 1986. Of the three Planning Advisory Group proposals outlined above only the second emerged unscathed. The single category for shops and services suggested by the Group was replaced by a three-fold distinction, namely prepared food class, retail class, and financial and professional services. The third idea of relaxations in residential areas was dropped although sub-division of non-dwelling premises was allowed. This could have an impact, for example, in the sub-division of redundant factories or warehouses (*Planning*, 1986b).

Eventually the new Town and Country Planning (Use Classes)

Order came into force on June 1st, 1987 and this new Order more or less followed the consultation document. Circular 13/87 (DoE, 1987c) sets out the purposes of the Order as:

1 To reduce the number of classes while retaining control over changes of use which, because of environmental consequences or relationship with other uses, need to be subject to specific planning applications.
2 To ensure the scope of each class is wide enough to take in changes of use which generally do not need to be subject to specific control.

The first of these purposes is interestingly expressed as it implies that the purpose of planning control is restricted to amenity and 'good neighbour' criteria. This is referred to by the Chief Planner of the London Planning Advisory Committee (LPAC, 1987, p. 2) when he suggests that Circular 13/87 argues that 'economic/employment or social considerations are not material considerations'. He says that if such criteria were considered, a different UCO would have been prepared. The restrictions created by the Order are particularly evident in the new business class, which combines offices and light industry. The combination means that it is impossible for local authorities to implement many of their employment-oriented policies. The LPAC Report looked at six local plans in London and in all cases identified at least three employment policies that would be affected, particularly those restricting office development to particular locations and those seeking to stimulate industrial uses.

Thus the new business class seems to reduce the power of the local authority. However, as noted, the Order rejects the Planning Advisory Group Report suggestion of a single retail class and this time the DoE seems to be supporting the local authority's ability to implement their retail policies. In rejecting this recommendation they say,

> many local authorities have policies designed to maintain and strengthen the retail element in primary areas, dominated by shops for the retail sale of goods. Adoption of either stage 3 or stage 4 of the sub-group's recommendations would take away the ability of local planning authorities to implement these planning policies.

(DoE, 1986a, para. 13)

Thus the government seems to be saying that local authorities need to retain their ability to implement retail policies whereas the power to enforce office and industrial policies can be removed.

The property advisers Debenham, Tewson & Chinnocks (1987)

have commented that the new Use Classes Order is likely to bring about increased opportunities for development. They identify two areas where this is most likely to arise. First, in London on the fringe of the City and West End where there is a high demand for office space. In these areas there are opportunities for the conversion of light industrial accommodation. The second kind of area is in the motorway corridors of the M4 and M25, where there is pressure as a result of the demand for out-of-town greenfield sites. The conversion of high-tech schemes to pure offices is likely to increase the value of such schemes.

Several commentators have pointed to the problems of definition regarding the new office categories (for example Titmus *et. al.*, 1987; Oppenheimers, 1987; Debenham, Tewson & Chinnocks, 1987). One of the problems here is deciding whether an office falls into class A2 or B1. The test for inclusion in the first category is that the office provides services principally to the visiting public. However, many offices do this as well as carrying out administrative functions and the Order does not specify any measures of the amount of public access required to fall into the A2 category. There is also the problem of flexibility, as such uses as banks and building societies vary the balance of their functions. A second problem arises with the new business class B1 which is designed to allow flexibility between office, light industry and research and development. However, a test is applied that all aspects of use must be capable of being 'carried out in any residential area without detriment to the amenity of that area by reason of noise, vibration, smell, fumes, smoke, soot, ash, dust or grit.' One problem here is that any office use that does not meet this test and does not fit into category A2 is not covered by the Order.

Some of the uncertainties of definition would seem to allow discretion and thus the outcome will partly depend on how far authorities take into account 'the spirit of the Order'. Examples of possible discretion occur in relation to whether permission is needed to turn unimplemented light industrial permission into office use or the way in which local authorities interpret what is a 'reasonable' condition. According to the report by Oppenheimers (1987), even if changes of use between the categories now covered by the new business class B1 do not require permission, operational development will still require permission. This could restore local authority control.

> In considering such an application, the ability to control the change of use within the new Class, e.g. B1, will be restored to the planning authority. It will, therefore, be open to them to refuse planning permission on the grounds of inadequate servicing, lack of car-parking spaces, design etc.
>
> (Oppenheimers, 1987, p. 38)

Most commentators agree that these difficulties of definition and the expectation that some local authorities will try and protect their local plan policies will inevitably lead to an increase in appeals. Notwithstanding the opportunities that loose drafting might present to local authorities, there is also general agreement that overall the Order does provide greater freedom for developers and a restriction on the ability to implement planning policy. As the Oppenheimers Report puts it, 'effective town planning (at least so far as commercial development is concerned), will be taken further away from the democratically elected bodies and placed in the hands of market forces' (1987, p. 45).

Conclusions

Although the modifications to the development control legislation have not been as extensive as those affecting development plans, development control has been significantly restricted. The changes will be examined in relation to the framework set out in Chapter Five. The market-led approach is illustrated in nearly all the initiatives examined in this chapter and in particular in the removal of constraints on industry and small business in Circulars 16/84, 14/85, 2/86, and in Circulars 9/80 and 22/80 where increased power is given to housebuilders. It is stated that the purpose of planning is to 'create the right conditions for economic growth and encouraging employment'. In Circular 13/87 on the Use Classes Order it is implied that any additional purpose of planning is restricted to amenity and 'good neighbour' issues. Many of the government's initiatives and statements on development control limit the scope of planning. The criterion of 'non-conforming use' is excluded from most considerations as a result of Circular 22/80 and development control now has less power over design matters. Changes to the GDO remove larger building extensions from control while the UCO modifications and the proposed 'flexible permissions' limit office and employment controls. Stricter limits are also applied to planning conditions and Section 52 agreements. These changes will make it difficult to implement policies aimed at office restriction or the protection of particular economic activities. Such policies are seen as interventions into the normal functioning of the market and therefore, from the neo-liberal perspective of Thatcherism, undesirable.

The appeal decisions since Circular 22/80 show how the demand for housing or industry often conflicts with environmental criteria. This kind of conflict has been minimised, to some extent, by excluding from the relaxation of control the most sensitive environments, such as conservation areas, National Parks and Areas of Outstanding

Natural Beauty. For example, Circular 22/80 and the changes to the GDO do not apply in these areas.

The second dimension of the framework was that of the **procedures** adopted in the planning system. Here there are again two clear trends: the increase in centralisation and the reduction in participation. Central control has the effect of reducing the ability of local authorities to introduce their own criteria and allows central government to ensure that market criteria dominate, as expressed in their increasing involvement in planning gain. Again, the reduction of opportunities to participate is a thread running through the period, associated with the desire to speed up and streamline the system. These reduced opportunities can be seen in the limitations on development control such as in Circular 22/80. Circular 2/87 (award of costs) clearly states that the views of local communities should be ignored unless they are 'founded upon valid planning reasons which are supported by substantial evidence' (para. 9; see also DoE, 1988a, para. 18). A further reflection of the authoritarian anti-democratic aspect of Thatcherism can be seen in the attitudes expressed on the issue of design. Heseltine, when Secretary of State, said that on the question of aesthetics the principle of democracy may have to be rejected. This is a view that is also strongly expressed by the neo-Conservative theorist, Scruton, who was mentioned in earlier chapters as a contributor to the authoritarian strand of Thatcherism (see Punter, 1986).

The net effect of the modifications to the development control system is to retain a strong planning system operating in certain areas where conservation and environmental factors are considered important. Elsewhere the system has been much modified and weakened and market criteria are expected to dominate with the removal of many previously adopted criteria. The public are expected to take a less involved role and central government has increased its surveillance to ensure that the market is given its freedom. There are also an increasing number of categories of development where previous planning controls have been removed altogether.

Chapter eight

By-passing the planning system

Modifying the existing planning system is not the only way of creating change. The Thatcher governments have also explored mechanisms to avoid or by-pass the normal system, based on the 1971 Town and Country Planning Act. They have experimented with architectural competitions, Special Development Orders and Urban Development Corporations. Most of this chapter is devoted to the last of these initiatives because Urban Development Corporations (UDCs) have become an enduring and growing aspect of the government's approach: Although the frequency of use and geographical coverage of the three initiatives may be restricted, they do have the potential for further extension and the attitudes they incorporate could have an indirect effect on the operation of the broader planning system.

Architectural competitions: an early idea

In a speech to the RIBA in July 1980 Heseltine expressed his interest in architectural competitions. He returned to the theme in a later speech to the British Property Federation in March of the following year. He urged the private sector to promote competitions, reported that he had written to over 100 chairmen of major companies and that there would be a number of forthcoming competitions in the public sector. The first competition with planning implications was announced later in the year by Arunbridge, the project managers of the Effra and Green Giant development site. Although the entries did not produce any outstanding schemes a winner was eventually chosen. However, the subsequent development of the site did not materialise as hoped. The competition for the National Gallery extension which was launched at the time was also fraught with difficulties. The concept seemed to lose its attraction to the government and faded from the scene; however, its consideration at the time raises some interesting points about attitudes to the planning process. The effect of the competition is that the normal process of planning permission with its links to local democracy and opportunities for objection is replaced by a system of judgement by an expert panel with no opportunities for wider discussion, participation or influence.

Special Development Orders: instruments of freedom

Another approach used by Heseltine when Secretary of State was the extended use of Special Development Orders. Again this initiative has the effect of by-passing the normal planning procedures. Section 24 of the Town and Country Planning Act 1971 gives the Secretary of State the power to make Special Development Orders (SDOs) which alter the controls or procedures in the General Development Order. The SDO has the effect of granting planning permission for a specific development or area of land and has to be agreed by Parliament. Until recently the SDO was used to increase controls in environmentally sensitive areas such as National Parks and in cases that merited a Parliamentary debate, such as the Windscale nuclear plant. However, in June 1981 Heseltine, responding to pressure from the industry and construction lobby, issued a consultation letter that proposed to extend the use of SDOs (Anderson *et al.*, 1981). The letter said 'the purpose of making fuller use of the SDOs would not be to make any general relaxation in development control, but to stimulate planned development in acceptable locations and speed up the planning process'. The reaction from the planning profession and local authorities was a fairly unanimous objection to the idea on the grounds that it would interfere with the responsibilities of local authorities and undermine local democracy. Reservations were also expressed on the reduction in participation opportunities as the SDO avoids the need for a public inquiry.

In May 1982 Heseltine proposed the use of SDOs to deal with the requirement of the Mercury Consortium to obtain planning permission for its telecommunication network. This proposal did not generate antagonism as the cables were to be laid alongside major railway routes and without the SDO would have involved separate planning application to each authority through which the network passed. However, the earlier (November 1981) proposal to use SDOs on the Effra and Green Giant site on the South Bank of the Thames did cause controversy. As mentioned above, this site was also the subject of an architectural competition and Heseltine proposed that the winning entry would get automatic planning permission if the developers decided to proceed. He described the South Bank SDO as 'the first experiment' and justified its use, in combination with the architectural competition, in the following terms: 'it enables a degree of involvement much earlier that might lead to a greater certainty for the developer if he works from the outset within a planning brief that has my support. It may also lead to useful savings in time' (quoted in *Planning*, 1981e). However, a typical response was that of John Finney, Junior Vice-President of the RTPI at the time, who said that the use of the SDO in such a site was wrong as it raised no issues of national importance yet local conflicts were considerable and the public would have no opportunity to discuss or influence the decision (reported in *Planning*, 1981e). Once the competition result was announced Heseltine proceeded to lay the SDO before Parliament. The Order was debated in the House of Commons at the end of June 1982 and passed by 132 votes to 86. In the debate the government spokesman, Sir George Young, denied that the SDO was undemocratic and claimed that the development was of more than local significance. The House of Lords subsequently also passed the Order. However, the SDO and architectural competition initiatives suffered a setback when it was announced in August 1983 that the project managers of the scheme, Arunbridge Ltd, had been forced into liquidation.

Even more controversial was the decision to apply the SDO to the Hays Wharf site in north Southwark. Under Section 148 of the Local Government, Planning and Land Act 1980, SDOs can be used in Urban Development Areas for giving planning permission to any development. The London Dockland Development Corporation (LDDC) had produced a brief for the site and the development company, St Martins, produced a scheme that exactly met the brief. This scheme was to be marketed as Europe's largest single urban renewal project. The LDDC then submitted the proposal to the Secretary of State, now Jenkin, for approval through an SDO. This mechanism excluded public consultation. This was done against a background of controversy over the use of the site. A previous scheme had been the subject of a local inquiry in which objections were made by the London Borough of Southwark and the local community groups. The objectors claimed that the office content in this earlier scheme was excessive and that housing and industry were needed – a view that was being embodied in the preparations being made at the time for the statutory local plan for the area. The inspector had supported these views and recommended refusal of the scheme. However, Heseltine did not accept the inspector's view but he did refuse part of the scheme on the grounds that the office developments were 'excessive and the scale and massing are inappropriate to its location'. It was therefore expected that a revised application would be submitted to take these views into account. Instead the scheme that was the subject of the negotiation with the LDDC and thus the subject of the

SDO had an increase of 10 per cent in its office content. Jenkin justified his agreement to the SDO in only two pages of A4. In this short report he stated that the use of Section 148 was particularly appropriate for the development of considerable tracts of land that had previously been used for purposes for which there was clearly no longer a demand. This suggests that the normal planning system is seen as inappropriate in derelict areas; a view that is reinforced by the designation of UDCs in such areas. The report also states that 'the granting of permission or approval for one particular form of development does not preclude the granting of planning permission (or the approval of development by some other statutory procedure) for other forms of development of the same site' (*Planning*, 1983d, p. 4). This attitude would seem to relate to the 'flexible frameworks' of the LDDC which will be discussed later. However, as MacDonald has suggested, the planning system becomes pretty meaningless if 'planning permissions are to be merely stepping stones to enable the market to find its own best use for sites' (1983a, p. 7).

Thus the process using the SDO ignores local interests and avoids the planning system which attempts to incorporate local needs and involve the public through participation. It has been reported in relation to the second scheme for Hays Wharf that 'at no time did St Martins or the LDDC make any plans available to the public, or send out notification to those living or working on or around the site, or hold any meetings' (Thompson, 1983, p. 11). This has led one commentator to conclude that the use of the SDO at Hays Wharf

> reveals how fragile the planning system is despite steady progress over the last twenty years to become more participatory and more open. Central government still has the power to act as if that system does not exist and as if the patient work of many to make the system more locally sensitive had never happened.
>
> (MacDonald, 1983b, p. 269)

Urban Development Corporations: marketing flexibility

Background and planning powers

Urban Development Corporations are the most important of the 'by-passing' initiatives especially in terms of their broader ideological impact. They played an important role in the 1987 election campaign when their application was extended. The idea was first adopted by Heseltine very soon after the 1979 election victory. In a press release on Inner Cities Policy issued on September 14, 1979 he announced the intention of setting up Urban Development Corporations in the London and Liverpool Docklands because in these areas there 'is a need for a single minded determination not possible for the local authorities concerned with their much broader responsibilities' (DoE Press Notice 390). This was quickly followed up by a consultation document sent to local authorities and other interested bodies (DoE, 1979), and the initiative was enshrined in legislation as Part XVI of the Local Government, Planning and Land Act 1980. This Act gives the Secretary of State the power to designate areas in metropolitan districts or Inner London Boroughs, subsequently extended to other kinds of authorities, as urban development areas. Having designated the area, an Urban Development Corporation then has to be set up. The designation of the area and the setting up of the Corporation are done through an order approved by each House of Parliament. The order for the Merseyside Urban Development Corporation was made in November 1980. The order for the London Urban Development Corporation was made at the same time but objections were lodged by local authorities and community groups in the area and a Select Committee of the House of Lords was set up to consider these. Before looking at the issues raised in this Committee the planning-related powers of the Corporations are outlined.

The Act sets out the general powers of the Urban Development Corporation but there is some flexibility for each order to vary the details. Most Urban Development Corporations are given full development control powers in their areas including control over listed buildings, conservation areas and tree preservation. There is some variation from this, for example the local authority can carry out the development control function on an agency basis or, as in the case of Cardiff, a special group of local authority, UDC and consultants can be set up (Stansfield, 1987). Whatever the arrangement the local authorities in the area lose control over the implementation of planning policy and the strategic authority loses its power of direction over major applications except those involving highways. In carrying out its development control function the Urban Development Corporation 'is required, like any other planning authority, to have regard to the provisions of the development plan and to other material considerations' (DoE, 1981d, para. 8). However, as outlined below the position regarding development plans in the designated area lacks clarity and in any case, as shown in Chapter Six, there is a general downgrading of the importance of plans with more attention being given to these other material considerations.

The situation regarding forward planning powers is complex. Formally the powers and duties of preparing development plans lies with the existing local authorities. As set out for the London Dockland Development Corporation in a DoE Memorandum, the Corporation is 'expected to take the existing statutory development plans and the London Dockland Strategic Plan as its starting point in formulating its own planning views and to take into account draft local plans and other informal plans prepared by the Boroughs' (DoE, 1981b, para. 4). However, as described further below, there is concern over the manner in which the LDDC moves from this 'starting point'. If a local authority believes that a proposal supported by the LDDC is a major departure from the statutory plan then the LDDC sends it to the Secretary of State for direction. Although the local authorities retain the plan-making role, the Urban Development Corporation can also submit proposals directly to the Secretary of State for the development of land in its area. The Secretary of State consults the local authorities and then may approve the proposals, possibly with modifications. Planning permission is then granted through a Special Development Order for any development that conforms with the proposals. The proposals themselves when submitted to the Secretary of State can be in a very general form and do not require the surveys, consultation and public participation needed under normal planning procedures. As far as participation is concerned the Urban Development Corporation is required to produce a code of practice setting out its means of consultation. Confusion over plan-making arises because although the local authorities are the formal plan-making bodies, the extent to which the Urban Development Corporation needs to conform to these plans is unclear and the Corporation has its own channels for formulating plans. As stated in a DoE Memorandum, 'the Corporation will make known its own planning views by issuing policy statements, planning briefs or plans for all or part of its area.... Local authorities will be expected to take these views into consideration in preparing any plan which affects the area' (DoE, 1981b).

The development control powers and plan-making abilities of the Urban Development Corporation have led one writer on planning law to conclude that 'the major consequence in planning terms of the reorganization of planning functions is that the new corporations subsume all the powers formerly exercised by the various authorities for the area at borough and district, and strategic level' (Grant, 1982, p. 543). Such a view is reinforced by a Chief Executive of the LDDC who has said that 'with land ownership and the planning powers vested in it by government, the LDDC can act as its own planning authority' (Ward, 1982, p. 14).

Land acquisition is another important power allocated to the Urban Development Corporation when considering the implementation of policy. The Urban Development Corporation has the 'power of assembling land, reclaiming it, servicing it and either developing the land itself, or disposing of it to private developers' (DoE, 1981a, para. 6). The Corporations have considerable powers of acquisition and can apply to the Secretary of State for a vesting order to transfer public sector land to the Corporation. Again this can result in major loss of control for local authorities who can find it extremely difficult to implement their policies as a result of these transfers. In the LDDC area 870 acres of public land was vested in the Corporation by 1983 and the Merseyside Corporation owned 75 per cent of its designated area by that time.

Before turning to the main issues raised by the Urban Development Corporation initiative it is necessary to note briefly the differences between the first two designated areas. The first major difference relates to the population living in the designated area. The Merseyside area contained only one small council estate whereas in 1981 there was a population of 39,700 in the London Docklands (GLC, 1984). The communities making up this population have a history of active involvement in local decision-making and thus their expressed interests and needs were bound to be a major issue in the life of the LDDC. A second major difference relates to the economic context in each case. In London Dockland, although the Docks have moved and manufacturing has declined, there are major pressures for other kinds of development such as commercial and private housing resulting from the proximity of the City. In Merseyside there is little private sector interest of any kind. The local authorities have reacted differently as a result. Whereas in London the authorities have, in the main, opposed the actions of the Corporation, in Merseyside after initial opposition from Merseyside County Council, the authorities have welcomed the activity of the Corporation, and the public money allocated, in the hope that investment could be attracted. In this case the aims of the local authorities and the Corporation are less likely to conflict. The view in Merseyside seems to be that the Corporation complements the local authorities and carries out works which the authorities would have undertaken themselves anyway (Boaden, 1982). This is reflected in the fact that the previous plans for the area have not been in conflict with the Corporation's strategy (see Adcock, 1984). However, in London, as Lawless (1986) points out, central government was determined to change the whole orientation and direction of local authority planning policy in the area.

The death of local democracy

One of the major issues raised by the powers allocated to the Urban Development Corporations is the effect on the processes of local democracy (Duncan and Goodwin, 1982). According to Ambrose (1986), the post-1947 planning system has been effectively by-passed in docklands since 1981. The area has been 'taken into care' by central government because its natural parents, the local boroughs, were too leftish, too committed to local needs and too sensitive to local feelings to carry out the kind of private sector led redevelopment strategy the Thatcher government had in mind.

(Ambrose, 1986, p. 251)

This comment illustrates the ideological dimension of the Urban Development Corporation initiative and a major area of debate it has generated, namely the lack of emphasis given to local needs and the removal of local democracy.

An early crystallisation of these issues occurred in the discussions in the House of Lords Select Committee set up to consider objections to the LDDC. The Committee was given the brief of judging whether the existing authorities were capable of regeneration or whether there was a need for the new authority. The discussion focused on the question of efficiency and the value of an organisation that can speak quickly with 'one voice'. However, the comparison between existing authorities and the Urban Development Corporation was not one of comparing like with like, as the finance and powers available to the new body exceeded those which were previously available to the authorities. Thus behind the surface debate on efficiency lay a more fundamental one of the process of decision-making. Should the needs of local people be overridden in the interest of some broader purpose? Should local democracy be by-passed? Should market criteria determine development at the expense of social criteria?

The anti-democratic nature of the Corporations, with their boards appointed by the Secretary of State, created adverse comment in both Liverpool and London. The Chair of Merseyside County Council described the Corporations as 'quangos of the first order, faceless, bureaucratic, and without a vestige of public accountability' (quoted in Lawless, 1986, p. 105). However, as mentioned above opposition in Liverpool diminished in the face of the extreme problems of economic decline. In London the antagonism persisted between the Corporation and local authorities and community groups in the area. The government stated its case for replacing local authorities to the House of Lords Committee. They claimed that the problem of regeneration of the area was a national one not a local one and that 'the boroughs tend to look too much to the past and too exclusively to the aspirations of the existing population and too little to the possibility of regenerating docklands by the introduction of new types of industry and new types of housing' (Select Committee of the House of

Lords, 1981, para. 6.4). The government suggested that the boroughs are subject to pressure from local organisations that are concerned with securing benefits for their members. They said that it is natural that such organisations should oppose the Urban Development Corporation because they realise that they will not be able to subject this new agency to the same pressures that they have been able to exert on locally elected councillors. Thus the government concluded that because of the local authorities' emphasis on the local community 'the boroughs are unsuitable recipients of money' (para. 6.7) intended to pump-prime the private sector. In their conclusions the Committee supported the government's views and suggested that some of the opposition was based upon a parochial attitude. They stressed, however, the need for the Corporation to gain the confidence of the local authorities and local organisations and the importance of the Code of Consultation in this respect.

Thus the regeneration of the two Dockland areas is organised by government-appointed Boards accountable only to the Secretary of State. The boards have the discretion to judge how much they inform the public or consult with local authorities and local organisations. In London these latter bodies are continually complaining about secrecy and lack of information. This arrangement is justified by the government on the grounds that the regeneration is in the national interest and thus the appropriate democratic body is Parliament and not the local authorities. However, as Colenutt (1981) points out, this national interest has not been defined. Martyn (1981) has claimed that the government's overt rejection of the legitimacy of local groups and their comments on the interaction between local councillors and their constituents represents a direct attack on local democracy and 'the importance of the balance between local and central government' (1981, p. 293). As he says, the government's attitude, that disagreements in the past between local authorities and the community demonstrates a failure of democracy, denies the fact that conflict is an inherent element of politics. McAuslan has neatly summarised the situation as one of replacing the development approach of a partnership between the local community, the local authority and central government with an approach of appointed officials developing a piece of real estate in the public or national interest (1981, p. 255).

McAuslan has also made some interesting comments on the Code of Consultation which the Lord's Select Committee thought so important. He describes the inclusion of this requirement in the legislation as being rather odd and suggests that it was a last-minute addition. He points out that there is no indication whether 'this is a legally binding and justifiable code or merely a statement of good intentions' (1981, p. 254). The Secretary of State is not involved in the preparation or approval of the code which McAuslan interprets as an expression of its lack of importance in the eyes of the government. There is also no necessity for the local authorities to agree with the code.

What has happened in practice regarding consultation since the legislation was enacted? In London the Corporation set up a Dockland Steering Group in 1982 to discuss the machinery and criteria for the allocation of LDDC funds to voluntary groups. Local authority officers and representatives of local groups attend this Group. There are also area sub-committees that consider applications for the funds. Councillors from the constituent Boroughs were also invited to sit on the Board of the Corporation. However, they have been reluctant to do so as they would be accountable to the Secretary of State rather than to their Boroughs. The Docklands Forum, a body made up of community groups, trade unions and business interests, carried out a survey of attitudes of its members to consultations with the LDDC. This showed considerable dissatisfaction (see GLC, 1984, p. 15). In particular there was concern over the informal officer-level contacts favoured by the LDDC rather than formal sessions with representatives and the public. It was felt that the LDDC approach led to a selective coverage and could promote divisions between groups. 'They felt that there was an "open and closed system" operating, in which friendly chats with officers gave an appearance of openness, while in a real sense the Boardroom door was closed to local people and local views' (GLC, 1984, p. 16; see also Docklands Consultative Committee, 1988).

The LDDC draft Code of Consultation sets out the following procedures:

(a) the LDDC will consult local authorities and neighbours adjacent to the site affected;
(b) individual organisations felt to have a special interest will be advised of the proposal;
(c) a weekly press release will be issued of applications received in the previous 7 days;
(d) 14 days will generally be allowed for consultation, with certain more major applications having 21 or 28 days for discussion;
(e) potential applicants are encouraged to discuss their proposals prior to submission; local authorities will be invited to these meetings when appropriate.

(GLC, 1984, p. 9).

These procedures were evidently devised with speed of decision-making in mind. They also give the LDDC considerable discretion in deciding when and with whom to consult. A major reaction to the Code has been the short time periods given to authorities and groups to respond to proposals and the incompatibility of these time periods with council and group meetings. There is a conflict between democratic time-scales and market-oriented decision-making. Even when the London Borough of Southwark adjusted its committee operations in 1986 to speed up their reactions to LDDC proposals the borough's views do not seem to have had much effect on the Corporation's decisions (Docklands Consultative Committee, 1988). In 1986 the LDDC eventually opened up its Planning Committee to the public. However, observers have no speaking rights and major decisions are taken at the confidential Board level and thus there is little opportunity to exert any influence.

The market reigns supreme

As a number of commentators have pointed out (for example, Colenutt, 1981; Newman and Mayo, 1981; Lawless, 1986), the reason for disregarding local democracy and limiting consultation in the London Docklands is to enable the Corporation to downgrade certain social criteria and to promote a different kind of development in the area. As Broakes, the first chairman of the LDDC, said, 'we are not a welfare association but a property based organisation offering good value' (quoted in Ambrose, 1986, p. 228). The previous plans had emphasised industry and public sector housing which, because of the decline in manufacturing industry nationally and the cuts in public housing finance had been slow to materialise. In contrast to this the Urban Development Corporation is promoting private housing, offices and leisure developments. This action by the Urban Development Corporation implies a shift to developments that have no direct or immediate benefit for existing residents and a strategy that is based upon bringing into the area new residents that can afford the expensive housing and who work in the service sector. If successful then by the time the Corporation is wound up the whole nature of local politics in the area will have changed. Meanwhile the local authorities still have the responsibility of catering for the social needs of their existing populations which extend beyond the boundaries of the Docklands area. One of the arguments of these authorities is that the land in the Docklands is needed to satisfy the needs of this wider population. Such wider provision remains the concern of the local authorities in a context of dwindling resources. For example it has been said that 'local government must be concerned with such issues as increasing racialism, vandalism, reduction in the standards of health and education, growing unemployment, rising crime rates, etc., even if central government can ignore them in the short term' (Newman and Mayo, 1981). The Urban Development Corporation is adopting the approach of concentrating on the narrow task of stimulating development via the market and leaving the broader responsibility to local authorities, or the longer term when it is hoped that the spin-off effects of this development will cater for other needs.

The objectives of the Urban Development Corporation are set out in very broad terms in the 1980 Act as:

> to secure the regeneration of its area, by bringing land and buildings into effective use, encouraging the development of existing and new industry and commerce, creating an attractive environment and ensuring that housing and social facilities are available to encourage people to live and work in the area.
>
> (Local Government, Planning and Land Act 1980, Section 136)

However, right from the outset it was clear that this regeneration was to be achieved through the attraction of private sector development and that the Urban Development Corporation's role was to provide the framework of confidence and infrastructure that the market needed. The first Chief Executive of the LDDC, Reg Ward, in his early public statements elaborated on this approach. In a talk in 1981 (Ward, 1981) he said that the LDDC's major job would be the international marketing of the prime sites; the 'Isle of Dogs' becomes a marketable commodity. Companies should be persuaded, using the USA and West Germany as examples, that it is possible to create profitable enterprise out of social needs and social regeneration. Ward sees the Corporation as 'a resource centre which can identify markets and initiate action.... We look at the market, generate enthusiasm among potential investors' (1982, p. 11).

According to Ward, firms and finance will be told that they have an almost free hand guaranteeable at least to the next general election. In this climate it is hoped that companies will be more likely to take risks. Ward does not have a high regard for planning. He said, 'there is no product of planning over the last 50 years that could not have been achieved by environmental control'. The long-term attitude to satisfying the social needs of the population is again clearly illustrated in the following statement by Ward where he is describing the approach of the Corporation as

> providing the initial resources so that key developments can take place in any given localised area; attracting further spin-off business initiatives around them, to be followed by further desired

developments to satisfy the needs of the local communities and, in time and given the initial impetus, to ensure that the economic prosperity of the area will be self-perpetuating.

(Ward, 1982, p. 11).

Lock (1987) has described the decision-making process implied by this market-led approach. He explains how the consultants Conran Roche viewed their job when commissioned by the LDDC to advise on a regeneration strategy for Greenland Dock. He illustrates clearly how market criteria precede any consideration of planning or policy objectives. One of the first jobs was to carry out a

> market appraisal to discover what could be viable here given that ... there had been no development interest in the area. We were clearly going to have to create a demand where none presently existed. This exercise was unconstrained by planning policy: we needed to know what was possible before applying the policy filter.
>
> (Lock, 1987, p. 12)

It should be pointed out that the London Borough of Southwark, who owned the land before it was vested in the LDDC, did have proposals for the area. The quote also raises the question of what this policy filter involved. This aspect is explored further in the next section when the planning approach of the Corporation is examined in detail.

The Merseyside Development Corporation (MDC) started with the same basic objectives, as set out in the legislation. However, as mentioned above, there are considerable differences in relation to size, population, land condition and economic pressures. The Initial Development Strategy for the Merseyside Development Corporation was published in August 1981 and extensive consultations were carried out with local authorities and community groups. This strategy concentrated on the engineering and infrastructure aspects and much of the Corporation's early life was spent acquiring land and implementing these works (see Adcock, 1984). As far as land-use proposals were concerned these were left intentionally flexible and did not involve any confrontation with the local authorities' existing plans. As Adcock puts it, 'the Strategy quite deliberately took a flexible approach to the precise use and management of land and buildings. Thus the MDC can readily accommodate a wide range of propositions from investors and other private sector institutions and developers' (1984, p. 277).

As Wray (1987) indicates this initial strategy was oriented towards industrial development. He reports how, as a result of weak demand for these industrial sites, the Corporation undertook a fundamental review of its approach in 1985. This review involved a switch to a tourism/leisure-led strategy including plans for a retail park. Thus the realities of the market have led to modifications of planning policy. As a result the match between previous local authority plans and the MDC strategy disappears and it becomes an interesting issue whether the good relations will continue. As Wray points out retail and leisure projects may result in a shift of resources within the region rather than a basic improvement in the regional economy. Adcock (1984) suggests that the tolerance of the loss of local political accountability depends upon the ability of the MDC to generate inner city regeneration over a wide area. It is therefore interesting from this point of view to see the Corporation also extending its activity from physical infrastructure to that of business development and training initiatives. These changes in strategy have prompted Wray to ask whether Merseyside needs a different kind of agency covering a wider geographical area and a wider spectrum of concerns to replace the Urban Development Corporation, which is confined to a small derelict area.

A new planning system – the London example

The implications for the planning system in the UDC area are now examined through a more detailed look at the LDDC. As already described one of the objectives of the LDDC has been to change the direction of the planning policies in the area. This has meant that the LDDC has had to find a way of ignoring previous plans and those being developed by some of the boroughs who, as mentioned above, retain the legal responsibility to prepare statutory plans. According to Ambrose (1986) the Corporation received direct instructions from the DoE to 'short circuit the planning procedure' (1986, p. 228). This seems to be supported by the statement from the LDDC Chief Executive that 'planning applications can be submitted to the Corporation with the assurance of speedy action. For the new streamlined approach is not governed by the traditional restraints regarding land use' (Ward, 1982, p. 11). What is this new approach?

On another occasion Ward has said that 'planners presume that they can regulate the market-place – and they can't ... [the need is] to be responsive to development pressure ... which requires a very flexible planning framework' (Ward, 1981). As outlined above, the LDDC has the power to formulate its own informal plans which it has tended to call 'area frameworks' and over the years it has produced a number of these. In 1984, after considerable pressure, the Corporation also released its Corporate Plan (LDDC, 1984a). The

Corporate Plan sets out the overall policies that guide the Corporation's activities and local area strategies. From these documents it is possible to build up a picture of what is meant by a 'flexible planning framework'. As already mentioned the LDDC is required to take statutory plans as a starting-point for its own approach and also to 'take into account' other plans of the boroughs. However, as the GLC (1984) has pointed out this allows plenty of freedom to ignore these plans and there is considerable evidence to suggest that this is what has happened.

The LDDC has stated that its 'local area strategies involve a judgement as to the extent to which existing plans remain a firm foundation for the Corporation's activities, enabling implementation to proceed on a site by site basis' (quoted in GLC, 1984, p. 10). The Corporation believes that 'a rigid, conventional master plan approach is not appropriate to Docklands. Nor is the market demand yet so strong that potential developers and customers can be automatically turned away if they do not meet strict, fixed criteria' (LDDC, 1984a, pp. 7–8). Thus the Corporation's development frameworks are merely for guidance and can be disgarded if alternative proposals are forthcoming. These frameworks and site-planning briefs provide guidance to institutions, developers and companies without imposing excessive rigidity which may deter development initiatives. They state the design and environmental criteria which the Corporation operates through its landlord and development control powers. However, development proposals which do not meet these criteria will not necessarily be ruled out. The preparation of these flexible informal plans has resulted in the by-passing of the statutory plans thus creating considerable conflict between the statutory borough process and the LDDC's flexible frameworks. The story of the North Southwark Plan is an example of such a conflict.

In September 1984 the Public Inquiry was held into the London Borough of Southwark's North Southwark Plan, which had been produced with great involvement of local community groups. This plan stressed the need for industrial jobs, public housing with gardens and a ban on more offices. Thus it ran directly counter to the LDDC's strategy of stimulating the market and the resultant use of the land for offices and luxury housing. The LDDC was the principal objector to the plan at the Inquiry and went to considerable expense to oppose Southwark. In its proof of evidence it sets out its reason for doing so in the following way:

> the Corporation regards the formulation of a flexible and realistic adopted North Southwark Plan as being important to its own task of securing the permanent regeneration of the Docklands Area,

and its representations at this Inquiry seek to amend or remove specific matters which inhibit this remit from Central Government

(LDDC, 1984b, part II, para. 1)

The Corporation disliked Southwark's Plan because, 'the Plan is expressed in dogmatic terms. Many of the concepts embodied in the Plan are outmoded with no recognition of current trends influencing the economy' (part II, para. 8).

The LDDC's argument at the Inquiry was that in the ten years of the plan, development pressures could change and that these could not be foreseen. A land-use allocation could prevent or delay the development of sites. 'The need to react quickly and flexibly to changing conditions is considered to be essential if such major sites are to be developed and the permanent regeneration of this part of Southwark is to be ensured' (part III, para. 3). They therefore proposed that all the sites with development potential that fell into the LDDC area should be designated what they termed 'Mixed Development Areas'. These areas could then be the subject of detailed studies producing flexible guidelines for development, presumably via the 'area frameworks' of the LDDC. Thus the effect of their suggested modifications would be to make the statutory plan meaningless for the Dockland Area, placing all the plan-making function in the hands of the LDDC and their area frameworks and briefs. In his report the Inspector supported the LDDC view but Southwark decided to press on regardless and adopt the plan. However, the plan was then called in by the Secretary of State in 1986.

A detailed example of the differing approaches of Southwark and the LDDC can be seen in the proposals for Greenland Dock that lies within the North Southwark Plan area. In its plan Southwark allocated the area for housing, which they intended to be public housing, open space and industry. However, the LDDC commissioned the consultants Conran Roche in July 1981 to produce a regeneration strategy for the area. At this point the area was removed from London Borough of Southwark ownership and vested in the LDDC. The consultants produced their final 'Framework for Development' in April 1983 recommending housing with water sports and a small amount of small-scale commercial uses. This framework was largely accepted by the LDDC. The differences in the two proposals became one of the issues at the North Southwark Local Plan Inquiry in 1984. The local authority in their Newsletter associated with the inquiry expressed the conflict in the following terms:

the Council's plans for Greenland Dock show clearly how the LDDC's interests conflict with the community. A planning application for new industrial units was turned down by the

LDDC and now has to go to appeal. Instead of new jobs, the LDDC want to kick out existing employers and use the land to build luxury housing way beyond the reach of local residents. The North Southwark Plan challenges their legal and moral right to behave in this way.

(Southwark London Borough, 1984a)

In their evidence at the Inquiry the LDDC stated the purpose of their framework for Greenland Dock as

to provide guidance to prospective developers on the character and disposition of the development components. It is also a strategy which the Corporation considers will continue to attract development interest, and so be viable in implementation terms within about a five year period.

(LDDC, 1984d, para. 1.4.9)

Thus the implication is that, in contrast, the Southwark plan is unrealistic with its emphasis on industry and public sector housing. They also claim that as the LDDC has started to implement its framework the statutory plan should incorporate this.

'The Corporation as landowner and development agency has thus commenced the implementation of its Framework for Development, and the London Borough of Southwark as plan making authority has, despite having full knowledge of the proposals, not taken them into account in the Plan'.

(LDDC, 1984d, para. 1.5.1)

In their evidence the LDDC then make detailed objections to the borough's proposals and suggest their replacement with the 'Mixed Development Area' allocation. Thus the resolution of the conflicting views is that the LDDC uses its power as landowner and development control agency to impose its views supported by its ability to produce informal strategies and plans. The statutory plan is then expected to conform to their approach.

In essence then the planning process in London Docklands is determined by the Corporation. The vesting of land into the ownership of the Corporation and their development control powers enable them to determine the future land use of the key development areas. The fact that the legal responsibility for statutory plan-making still lies with the boroughs does not prevent the Corporation from taking the leading role. As the Southwark example shows, they can still get their way even when the borough has different objectives. This conflict might cause some inconvenience, such as having to spend time objecting at a Local Plan Inquiry, but has had no effect on

178

changing their plans. Thus the ability to prepare informal plans seems to be sufficient for their purposes. However, some people have argued that the LDDC should be given plan-making powers, for example David Lock has said that 'I think UDCs should have statutory planning powers to remove procedural loops created by hostile authorities and to cultivate a more public planning process' (Lock, 1987, p. 15).

In practice therefore the development frameworks have replaced the statutory plan-making process as the vehicle for determining future development. A major feature of the frameworks is their flexibility and their main purpose is to generate market interest and confidence in sites rather than control exact land allocations. Thus if other commercial propositions are forthcoming then the framework is set aside. The only controls are at the detailed level of development control criteria. Although some consultation takes place around the development frameworks this is far less rigorous and extensive than that carried out on statutory plans. The flexibility mentioned above will also diminish the value of these consultations as changes take place and are agreed in meetings between developers and the Corporation. Another change brought about by the UDCs is the amount of plan-making undertaken by private planning consultants who have very weak connections with the local community.

The spread of corporate control

It has often been suggested in Conservative circles that the UDC initiative should be extended. For example, after the Tottenham uprising in 1985 it was proposed that the UDC concept should be applied to such riot-prone areas. In this suggestion it was explicitly stated that the main advantage of doing so would be to by-pass the decision-making processes of local authorities such as Haringey that were seen to be pursuing detrimental policies. Thus local democracy would be replaced by a centrally controlled agency ensuring that the 'correct' approach was taken.

However, nothing came of this particular suggestion and the extension of the concept occurred later in the preparations for the 1987 election, when further designations were suggested and included as a major element in the Conservative Manifesto's inner city policy. The five new UDCs were Cardiff, Trafford Park, Tyne and Wear, Teesside and the Black Country and these were put into effect in the spring of 1987. The response of local authorities was, on the whole, not antagonistic although there was some opposition in the Midlands, which led to a slimming down of the UDC area. After debate politicians in these new UDC areas decided that accepting and co-operating with the new Corporations was the only way of getting

179

government money into their areas (Colenutt, 1987; *Planning*, 1987).

In December 1987 Ridley, Secretary of State for the Environment, announced the government's intention to set up three more, smaller UDCs in Bristol, Leeds and Central Manchester and to extend the Black Country UDC to include part of Wolverhampton. Then in March 1988 Mrs Thatcher launched her much-heralded *Action for Cities*, which included a further UDC in the Lower Don Valley of Sheffield and an extension of the Merseyside Development Corporation (Cabinet Office, 1988). Thus by 1988 there were eleven UDCs in England and Wales, indicating a major expansion of the initiative.

However, the new proposals have generated some criticism (see for example, Brookes, 1989; Thomas, 1989; Parker and Oatley, 1989). The Bristol proposal is particularly interesting because the urban development area contains very little derelict land. This dereliction had been the main feature and *raison d'être* of the earlier UDCs. Bristol City Council took objection to the designation and set out to refute the arguments put forward by the DoE to justify it. According to the Council the proposal was not based on local knowledge, was not needed for the regeneration of the area given Bristol's buoyant economy and had the effect of replacing the local democratic processes which are responsive to community needs with the desires of the property market. The City's objections were investigated in a Select Committee of the House of Lords, which confirmed the appropriateness of the UDC but did reduce the development area, including the removal of key sites next to the city's shopping centre.

As mentioned earlier the experience of the first two UDCs differs greatly. The LDDC managed to attract private investment at the expense of considerable local opposition while Merseyside found it difficult to obtain private sector interest but established better local co-operation. The kind of land use already in the Urban Development Area and the development pressures resulting from its location are the key variables. In this respect the new areas are expected to fall somewhere between the two and contain some variation, for example there is expected to be some development pressure in Trafford Park but none in Teesside (Duffy, 1987). The extension of the initiative brings to further areas the reduction in local democratic control and an increase in areas where the normal planning procedures are by-passed.

As mentioned above, the Merseyside Development Corporation, operating in its limited area, needs to ensure that the benefits it generates spread to the rest of the region in order for the Corporation to maintain its support and fulfil its broader rejuvenation objective (Adcock, 1984). It has also been seen in the London case that when an area includes an existing community then the effect of by-passing local democracy is increasingly problematic. Thus the fortunes of the other UDCs are likely to depend upon whether the strategy is to confine attention to a small derelict area, in which case any broader economic impact is limited, or to undertake a more ambitious strategy, in which case the democratic antagonisms are likely to be greater. The new UDCs have been heralded as a solution to the inner city problem but as such they would seem to be limited. As Owen Luder (1986) expresses it,

> both London and Merseyside Corporations control large, cohesive and clearly identified land areas, which could be taken out of the normal local government planning systems and operated separately by a non-elected and non-political body. Elsewhere, inner city problems are peppered around urban areas in the cities and large towns.
>
> (Luder, 1986, p. 23)

Although it is not clear how well the UDCs will succeed in dealing with inner city problems, one aspect *is* clear – the government's desire to take decision-making away from local authorities. In looking back at the designation of the LDDC, Heseltine has made no bones of this fact. He has said, 'we took their powers away from them because they were making such a mess of it. They are the people who have got it all wrong. They had advisory committees, planning committees, inter-relating committees and even discussion committees – but nothing happened'. In contrast, 'UDCs do things. More to the point they can be seen to do things and they are free from the inevitable delays of the democratic process'. Not only is the process speeded up but also it is possible to ensure that the large amounts of central finance are spent in the way in which central government think best rather than influenced by local interest groups and other pressures acting on councillors. In effect this means spending more money on infrastructure to make the areas more attractive to the market and less on more direct 'people-oriented' social projects. However, one of the dangers of this strategy is the alienation of the local communities as social projects diminish alongside the loss of local democracy. Urban uprisings in the 1980s show that this alienation is a real threat. It is also possible that if the UDC is successful and land values and house prices increase in its area then social tensions with the older community will also increase. As Duffy (1987) has remarked in relation to the LDDC, 'so great has been the house price explosion, [however], that one of the major aims of social policy – which is to reduce the potential for social unrest – seems largely to have been overlooked'. However, it is possible to regard this issue of social unrest as a short-term problem. If the conflicts created by the loss of democracy and initial extreme social contrasts can be contained then, in the longer term, stability will be restored through the radical changes in the social composition of the area. Some commentators see this social reorganization as an important part of the government's strategy.

Conclusions

This chapter has examined three initiatives that by-pass the 'normal' planning procedures: architectural competitions, SDOs and UDCs. They each establish alternative mechanisms for dealing with land-use and development decisions. These alternatives all aim to 'streamline' the process and as a result involve reduced opportunities for participation. The UDC is the most important of these initiatives as it has become the most widely used. UDCs are given development control powers thus creating a potential tension between the policy-making process in local authorities and implementation carried out by the Development Corporation. The powers to vest land in a Development Corporation give further strength to their implementation capabilities. It has been seen that although the formal plan-making function remains with the local authorities the Corporations have the ability to override these plans if they do not conform to their wishes. It can therefore be said, as indeed the LDDC Chief Executive has claimed, that the UDCs have their own planning system. This system operates with very clear objectives. The aim is to generate market confidence in the area through stimulating those activities for which there is a market demand and providing for their infrastructural needs. This implies less public involvement and less concern for social needs. The acceptance of UDCs by local authorities, because of the expected economic benefits, thereby also implies an acceptance of a shift in the objectives and priorities of the planning system. It could be said that the eleven UDCs cover only a small geographical area and hence are not all that important overall. However, their impact could extend over a larger area. The ability of the local authorities, in whose area the UDC falls, to plan strategically is severely handicapped as many potential development sites are removed from their control. This therefore has an impact throughout the whole authority. It might also be said, although this is difficult to demonstrate precisely, that the operation of the UDCs has a wider ideological impact and affects the general balance in the relationship between planning authorities and developers, particularly in authorities that see themselves in competition with UDCs for development.

Turning to the framework set out earlier in the book, it is clear that the alternative planning processes adopted in these 'by-passing' initiatives, imply a shift in the principles of decision-making. The discussion of the LDDC and the views expressed by its Chief Executive have shown the very dominant role of the market. This is clear, for example, in the account of the consultants' planning process for the Greenland Dock in which market scope was assessed first, and only then was it considered how planning policy might fit in. Similarly in Merseyside once the industrial strategy embodied in the plans for the area was seen to be out of line with market demand, then the plan was abandoned for more market-led activities. A constant theme in all these discussions has been the need for planning to be flexible to the rapid changes in market demand and to ensure that there is a body, in the form of the UDC, that has the 'single-minded determination' to respond quickly to these demands. This dominance of market criteria results in the downgrading of social criteria. In a situation of shortage of development land the conflict between the different criteria becomes particularly acute as demonstrated in the case study of the North Southwark Plan. The strategy adopted by the UDCs is that the social criteria will eventually be met through the process of spin off. However, the market-led development creates changes in the whole social composition of the areas. This could be described as 'social engineering' of a kind and can be contrasted to the social objectives that authorities have been trying to achieve in their plans and which, as shown in Chapter Six, have now been declared inappropriate.

The second aspect of the framework concentrates on the procedures of decision-making and again it is clear that major changes are involved. The idea of architectural competitions would replace the planning process and its links to local democracy with decisions by an expert panel. Both SDOs and UDCs give much greater power to central government, for example the Secretary of State appoints the Boards of UDCs, determines their finance and is the arbiter of any conflicts (note the call-in of the North Southwark Plan). Central government uses this power to by-pass local democracy and allows the UDCs to negotiate directly with developers. The democratic safeguard in this process is supposed to be the Code of Consultation that UDCs have to draw up. However, it is clear that this is not given much importance and that decisions are made in secret without much consultation with local authorities or local groups. It could be said that a redefinition of politics is occurring here. Local level democracy is seen as detrimental to development which is equated with the national interest. Hence local politics is described as parochial and self-interested and, because of the lobbying of local councillors, creating unsatisfactory effects on the decision-making process. Democracy should, from this viewpoint, be confined to the national level and central government is portrayed as the guardian of this broader interest. Crucial to this argument is the claim that development is in the national interest and thus warranting the dismissal of local politics. However, this claim is not elaborated in detail and generally rests on the argument that development generates economic progress which is essential for national recovery.

The final point worth mentioning here is the ideological preference, identified in earlier chapters, for a 'preconceived general framework'. It can be seen that the initiatives presented here all result in a move in this direction. Once someone has won the architectural competition this person knows exactly where they stand and no further procedural hoops or pressures exist. The SDO has the effect of granting planning permission without the need to go through the process of application with the attendant uncertainties of bureaucratic and local political involvement. In UDCs the developer can discuss ideas with the UDC in the light of the area frameworks which set out beforehand the environmental and design criteria. Thereafter they can assume the certainty of their development as the UDC has the development control powers. Again there is no further bureaucratic or political involvement.

Chapter nine

Towards a simplified planning system

Urban Development Corporations operate what is virtually an alternative planning system. The Thatcher governments have also been developing another idea which can be seen as an even more explicit alternative to the normal planning system. This idea is the simplified planning regime which contains a new framework of laws and procedures to condition the scope and operation of planning. This chapter looks at the development of this idea from its inception as part of the Enterprise Zone (EZ) to its evolution in Simplified Planning Zones (SPZs). The first of these initiatives, the EZ, has been applied only to selected areas and has been described as experimental. However, it has had a broader function in changing the climate of opinion and the simplified planning regime has been seen as having the potential for wider application. The instigation of Simplified Planning Zones was a step in this direction. The purpose of this chapter is to explore the degree to which these ideas reflect the ideological principles of Thatcherism and in particular the emphasis on market processes and the preference, drawing on Hayek, for general a priori laws to replace the existing bureaucratic system of discretionary intervention.

The Enterprise Zone idea

Origins of the EZ proposal – Peter Hall's contribution

Most writers on the origins of the EZ concept point to the influence that Professor Peter Hall had on Sir Geoffrey Howe, the political instigator of the policy (see for example, Anderson, 1980; Bond, 1980; Williams and Butler, 1982). As far as the planning aspects are concerned, frequent mention is made of the similarities of the planning schemes in the EZs to the ideas in 'Non-Plan' of which Hall was a co-author (Banham *et al.*, 1969). Thus it is possible to see connections with the ideas set out in Chapter Five (it is interesting to note that

S.M. Butler, 1981, traces the origins of the EZ idea to Jacobs). As mentioned in this earlier chapter, the 'Non-Plan' article proposed 'experiments in freedom'. These ideas stemmed from a frustration with the planning process and its drag on innovative and exciting environments but, as reflects its origins in a period of growth, the areas proposed for the application of the 'freedom' experiments were located in open country (for example, East Midlands, north of Nottingham, Nuthampstead, South Hampshire and the Isle of Wight). The article specifically excludes London because of the complexities of existing urban areas.

When Howe first launched his EZ idea he acknowledged the influence of Hall. He said, 'I was delighted last year to discover that a distinguished Socialist, Professor Peter Hall, was beginning to reach for the same prescription as myself' (Howe, 1978, p. 11). Later the connection between Hall's ideas and the Howe initiative was given further credence when, in the House of Commons, Howe again specifically mentioned his debt to Hall. However, the exact influence referred to here was not the Non-plan article but a speech by Hall in 1977 called 'Green Fields and Grey Areas' which was directed towards dealing with the inner city problem (Hall, 1977). In this speech Hall's attitude to the lifting of planning restrictions is not as straightforward as in the earlier 'Non-plan' article. His main attention this time is focused on finding new sources of innovation and enterprise for inner city areas. It would seem that many of his proposals to deal with this problem are compatible with a certain amount of planning intervention, for example, attractively planned industrial parks, creation of high amenity environments, high income housing schemes, urban motorways, tourist developments and theme parks. In fact he concluded the speech by saying, 'We need to *plan* for an *orderly* reorientation of our space economy' (P. Hall, 1977, p. 6, emphasis added).

However, Hall also described, in the same speech, the 'Freeport solution' which he called 'essentially an essay in non-plan' (1977, p. 5). This he referred to as a last-ditch solution to be used on a small scale in exceptional areas such as derelict inner city areas with little or no population. Again Hall does not make his approach to planning controls in these areas clear. Such controls are not specifically mentioned, but relaxation could be inferred from his comments that there should be 'minimum control' and that 'bureaucracy would be kept to an absolute minimum' (1977, p. 5). The relaxations that are specifically mentioned cover exchange controls, taxation, social services, industrial regulations, and trade union closed shop practices.

The only other mention of planning regulations in the speech was his support for the recommendations of the Bolton Committee on

Small Firms which reported in 1971 and which suggested that:

1 small firms be allowed to build up to 10,000 sq.ft. without planning permission;
2 local authorities should use powers to develop small-scale industry in residential areas;
3 alternative accommodation and compensation should be provided for small firms in clearance schemes.

In his subsequent defence of his ideas Hall has been keen to demonstrate the differences between his concept and the one implemented by the government (see P. Hall, 1982, 1983a). However, these contributions do not help to illuminate his attitude towards planning controls *per se* as he gives this subject little attention. As will be shown later, the EZ planning schemes are not completely 'non-plan' as they contain reserved sub-zones, specific uses requiring permission and the retention of environment and safety regulations. The impression gained is that Hall sees this later government approach as a watering down of the 'freedoms' in his Freeport idea (P. Hall, 1983a, p. 461). On the other hand in these more recent articles he also stresses that deregulation needs to be accompanied by certain state actions such as provision of venture capital and training. He also refers to the need for government provision of infrastructure in the form of workshop space, roads and services. This would presumably require some planning.

Hall's contribution to the debate over the relaxation of planning controls can be summarised as follows. In the late 1960s he, with others, advocated experimenting with areas of non-planning. He considered that this would be welcome to free up innovative developments and reduce unnecessary bureaucratic restriction. He advocated experimental areas in open country that could be exploited for their growth potential but considered existing urban areas such as London unsuitable because of their complex problems. Later in turning to the problem of inner city decline he advocated the pursuit of economic growth activities in these areas and suggested that planning should seek to encourage and maximise these activities, such as through the provision of industrial parks, high-tech centres, and urban motorways. However, at the same time he suggested that Freeports should be designated in extreme cases and on a small scale. These Freeports would employ a fuller relaxation of controls which would presumably include planning controls. More recently, he has pointed out that even in these areas a certain amount of state intervention would be necessary including the provision of premises and infrastructure. It has also been pointed out by Schiffer (1983) that

Hong Kong, which is a model for Hall's Freeport concept, has experienced much state intervention, such as the provision of social services, roads, land reclamation and other infrastructure. In this case intervention is paid for out of what is effectively a land nationalisation policy.

Howe's initial statement

In his famous speech at the Waterman's Arms in the Isle of Dogs, Sir Geoffrey Howe put forward the idea of EZs aimed at giving enterprise more freedom to 'make profits and create jobs' in the 'worst afflicted urban areas' (Howe, 1978, p. 1). In his account of the failure of previous policy to deal with the problems of urban areas and national economic decline, he made a number of comments on the effects of planning. For example he said:

> While whole communities have been virtually blitzed by 'planning' and stagnation, whole industries have fallen off the edge of the economic table.
>
> (Howe, 1978, p. 4)

> For a quarter of a century, the centre of my home town in South Wales saw its commercial life ebbing away while 'plan' succeeded 'plan' ... even the prosperity of county communities is now being stifled by structure plans which set out ruthlessly to limit future commercial activity to businesses that are already based within the county.
>
> (Howe, 1978, p. 6)

Against this background he advocated 'liberalising the entire planning system' (1978, p. 10) in order to promote economic incentive. He then proceeded to outline his EZ idea. The first key element he described was the removal of planning controls:

> First, planning control of any detailed kind would cease to apply. Any building that complied with very basic anti-pollution, health and safety standards and that was not over a stated maximum height would be permissible, for any lawful purpose. IDCs and ODPs would not be required.
>
> (Howe, 1978, p. 1)

He also went on to describe how the management of the EZ would require 'a new model of authority with some of the qualities of a New Town Corporation' (1978, p. 15) and which would be responsible for disposing of property for private development. This he claimed was necessary because of the difficulties in obtaining a common view between different authorities.

He saw planning as a break on economic initiative and advocated in the zones very minimal controls confined to pollution, health and safety standards and height of buildings. He also saw the need to supplant traditional democratic authorities in order to achieve success in these zones. This disdain for democratic authority is illustrated in part of the speech where he compared the problem of overcoming municipal inertia to that of the abolition of the monasteries by Henry VIII. He said that EZs can give a new lease of life to moribund urban areas just as 'many medieval communities certainly gained a new lease of life from the dispersal of monastic property to freebooting individuals'. He stated his belief that 'just as zealously as the monks and abbots strove to do the will of God, today's chief executives and planning officers seek to serve the will of democracy – but alas, with less fruitful results' (1978, p. 8).

Thus Howe sees planning as having a detrimental effect on society in blitzing communities and preventing economic innovation. The restrictions and grand designs of planning have to be removed. He also sees planners attempt to pursue the 'will of democracy', perhaps employing the concept of 'public interest', as misguided. We have come across these ideas in earlier chapters in the work of people like Hayek and Jacobs.

The ideological importance of the EZ idea

It is necessary to pursue these original ideas at two levels. First at the broad ideological level, that is, the function of EZs in changing attitudes and principles. If a change could be achieved at this level then the door would be open for the smooth implementation of the scheme and its eventual extension to other areas. Second, at the level of the specific implementation of the original idea, in terms of legislation and procedures. The precise form that this implementation takes will be the result of discussion, debate and modification of the policy. As Howe put it at the end of his speech, the idea could be modified by 'the grey men whose job it is to consider the "administrative difficulties" of any idea [and who] would be ready enough to start manufacturing the small print that could stop the initiative in its tracks' (1978, p. 16).

Howe's speech was made at a crucial moment in the history of Thatcherism. In the lead-up to the 1979 election the Conservative Party was keen to demonstrate its new dynamic approach and saw itself as spearheading the battle for ideas. As described in earlier chapters, this was encapsulated in such phrases as 'rolling back the frontiers of the state' and 'breaking with the post-war consensus'. Much of the electoral rhetoric and early government statements were concerned with moving the climate of opinion towards a greater acceptance of private enterprise. This campaign can be seen as the first stage in a process which is then followed by experimental policies conforming to this new ideology and subsequently by widespread application. Eventually this process can result in the replacement of earlier laws, procedures and practices. In this context it is interesting to note one of Howe's comments in his Isle of Dogs speech. He refers to his vision of 'communities queuing up to apply for EZ status' and states that if this could be achieved then 'we shall have gone a long way towards winning the debate' (1978, p. 15). As will be seen later Howe's use of the EZ ideas as an ideological attack to 'liberalise the entire planning system' was to be extended through the White Paper *Lifting the Burden* (DoE, 1985a) and the Simplified Planning Zones.

A number of writers have commented on the broader ideological function of the EZ concept. Massey (1982) points to the importance of 'the political and ideological function which the debate over enterprise zones performed', and she concludes that 'the main impact of enterprise zones has been ideological' (1982, p. 433). She describes how the launch of the EZ initiative played a leading role in the more general ideological onslaught against regulations and practices that interfered with the freedom of enterprise. Thus the acceptance of EZs by Labour local authorities and the implementation of the EZs has altered the balance, she claims, between the interests of industry and the interests of working people and local communities. Once this balance has shifted she believes it becomes more difficult to promote arguments for democratic control, environmental protection, resources for social services and public intervention generally.

Anderson draws similar conclusions and believes that 'Sir Geoffrey [thus] had the specifically ideological objective of changing the "state of the political argument"' (1983, p. 328). He suggests that Labour authorities in bidding for the funds available in the EZ programme have to accept a non-interventionist philosophy and diminish their socialist objectives and that this divides the Labour Party against itself. The process in EZs could also lead to a weakening of local democracy as decision-making is shifted towards both central government and the market. Anderson's view is that the Thatcher government has sought to undermine local authorities as they are often the advocates of policies that run counter to central government's national economic stance of keeping wages down and promoting economic efficiency through competition. Thus he sees the EZ policy as having a broader ideological importance that overrides the purely economic effects. As he puts it, 'although not very significant in strictly economic terms, they [EZs] are important in the continuing debate over the role of local authority planning and local–central government relationships' (Anderson, 1983, p. 316).

Another example of Howe 'winning the debate' is given by Shutt (1984). He also sees EZs performing a wider ideological function and points in particular to its effect in 'attacking local democracy and legitimating broader calls for lower industrial rates' (1984, p. 19). He believes that the rate exemptions in the EZs have made it easier for the government to campaign for lower rates for industry elsewhere and also to pursue its rate-capping legislation. He also describes how the local plan with all its public consultation has been ignored in the Salford EZ and he contrasts this 'normal' planning system with the very limited participation in the EZ planning scheme. (For further contributions on the ideological importance of EZs see Botham and Lloyd, 1983; Keating, 1981.)

Thus regardless of the degree of economic success of EZs, whether measured in terms of new jobs, new development, improved environment, contribution to the inner city or regional problem, it can be argued that the function of the initiative at the broad ideological level needs consideration. What contribution does it make to the ideological thrust of Thatcherism outlined in earlier chapters? It has been suggested by the writers mentioned above that this ideological function can be seen to reinforce two, linked, aspects: first, the greater freedom of action for economic interests through the removal of state regulations and second, less control over decision-making by the general public and the state at the local level.

Formulating the new planning regime

The role of the state in EZs

Before turning to the question of planning regulations in the EZs, brief mention will be made of the role of the state, generally, in this initiative. Have EZs resulted in a significant withdrawal of state intervention? Many commentators have shown that the net effect of the EZ policy has been increased state intervention in the designated areas as far as financial subsidy, promotion, and provision of infrastructure are concerned. Keating contrasts the rhetoric with the practice in claiming that 'from being an experiment in 'non-government' they have become an exercise in government intervention and public subsidy' (1981, p. 669). Similarly, Anderson concludes that there is 'more state involvement in the zones than in the economy generally' (1983, p. 342).

These conclusions are based upon the experience of EZ areas where local authorities have been heavily involved in the preparation and servicing of land. For example Shutt (1984) points out that in Salford progress on the EZ did not occur until £3 million of public funds were spent on infrastructure over and above the EZ incentives. (For further evidence of public expenditure in the EZs see Tym, 1984.) Such allocations of public money are likely to force local authorities themselves into a shift of resources and priorities, for example, in Scunthorpe, Barnes and Preston (1985) have suggested that resources allocated to the EZ have diverted attention away from problems of industrial dereliction elsewhere in the authority area.

In addition to directing local authority resources from other areas, the EZ policy also leads to a shift in the costs of providing infrastructure from the private sector to the authority. As Purton and Douglas (1982) point out:

> Under normal planning procedures a developer and a local planning authority frequently agree on the responsibility for the provision of infrastructure. (Section 52 Town and Country Planning Act 1971) If a majority of developments are automatically deemed to be granted permission there is no pressure on the developer to provide the necessary infrastructure. If this advantage is lost the responsibility and cost of such provision will be left to local authorities.
>
> (Purton and Douglas, 1982, p. 421)

(For additional examples of the transfer of costs from the private to public sector see O'Dowd and Rolston, 1985; and Williams and Butler, 1982.)

Another form of state intervention occurs through the subsidy of the tax and rate concessions. Thus the state can be said to be fairly active in EZs through subsidy, steering local authority resources into these areas and removing the burden of infrastructure costs from the private sector on to the state. Anderson summarises what is happening as a continuation of state help for private capital while at the same time using anti-state intervention arguments to remove those particular regulations and controls that could benefit labour. Thus, in line with the argument put forward by Gamble (1988), there could be an increase in state intervention in ways that aid the market, such as providing infrastructure while at the same time using the ideological rhetoric to remove intervention in other areas. It would seem from the strong attack on planning made by Howe in the Isle of Dogs speech that a key element of this reorientation of intervention is the removal of planning regulations. How far has this apparent aim been carried through in the formulation of the legislation and procedures? This question will be examined under two headings: first, the change in the *scope* of planning, and second, the change in the *process* of planning.

EZ planning schemes: changes in the scope of planning

Notwithstanding Howe's desire to avoid the interference of the 'grey men', there has been considerable modification of the idea as it progressed towards legislation. As described by Taylor (1981), the task of formalising Howe's proposals was undertaken by the Industrial Policy Group at the Treasury who consulted with other departments. At an early stage some aspects were dropped, such as the relaxation of health and safety and employment protection. It was decided to give the responsibility of administering the EZs to local authorities. The allocation of responsibility for the EZs to local authorities meant that overall co-ordination went to the Department of the Environment. Pressure was put on the DoE to act quickly in time for the 1980 Budget and the necessary legislation was tacked on as an extra clause and schedule to the Local Government Planning and Land (No 2) Bill. This had already had its second reading in the Commons and was being considered in Standing Committee. Only a few weeks were available for interested groups to comment and the Local Government, Planning and Land Bill passed through the Commons in the summer of 1980 with only minor amendments to the EZ schedule.

The process of firming up the EZ idea involved negotiation between the DoE and the local authorities. Within these negotiations, as Taylor points out, 'the main battle ground ... was over the relaxation of planning regimes in the EZs' (1981, p. 432). As shown in an internal DoE report (DoE, 1983c) this battle consisted of the DoE striving to keep the planning schemes simple while the LAs wanted to include many exceptions to deregulation. The battle was not only to take place in determining the particulars of each EZ but also being fought in the continued passage of the legislation itself. The Association of County Councils and the Association of District Councils sponsored an amendment to the Bill during its committee stage in the House of Lords on October 15th, 1980. This successful amendment gave local authorities (LAs) the right to reserve planning permission for certain classes of development subject to the agreement of the Secretary of State. This meant that control could still be exercised over noise, pollution, hazardous substances, nuclear installations and explosives, and particularly sensitive parts of the zone.

Many interest groups also wanted control over retail developments but the DoE, at the time, was inflexible over this issue. The Association of Metropolitan Authorities took up the cause and proposed an amendment, through their spokesman Lord Ponsonby, that major retail developments should still be subject to planning control. This was considered in the third reading of the Bill in the House of Lords on November 5th, 1980 and rejected. However, pressure was kept up and a campaign conducted by local Chambers of Trade. Eventually the DoE climbed down (see Butler, 1981). During negotiations in early 1981 the DoE allowed control over retail developments in the EZ planning schemes although the precise limits varied between EZs.

Thus in the passage of the Bill through Parliament and in the negotiations between the DoE and LAs, certain concessions were made to the original intention to deregulate planning. These exceptions to deregulation can be categorised into three kinds:

1 'externalities' – that is activities that have a very severe effect on neighbours, for example, noise, pollution, nuclear fallout and explosion;
2 'sensitive areas';
3 'competition' – in the specific case of large retail development it is accepted that planning still has a role in mediating the effects of 'free competition'.

An important aspect of all these concessions is that they can be applied only after approval by the Secretary of State. Apart from these concessions, planning controls would be lifted and as a result detailed design matters would be left to the developer (see DoE, 1983d).

EZ planning schemes: changes to the process of planning

The discussion of the EZ planning scheme so far has concentrated on the scope of the planning function. The implications of the initiative on the procedural aspects are now examined. One of the questions addressed is whether the planning schemes can be said to reflect the authoritarian strand of Thatcherism, for example, less local democracy, less participation and greater centralisation.

Hall has little confidence in local authorities and their ability to solve inner city problems. He is quoted as saying, 'I think that enterprise zones should have been taken out of local authority hands altogether' (Bond, 1980, p. 7). He is also said to be a firm believer in Urban Development Corporations and thinks that the 'loss of democracy is a sacrifice that must be made for the sake of the urban economy'. This is not of course surprising given the inspiration he draws from such places as Hong Kong and Singapore. He defends his view with the following argument:

> Many of these industrialised countries have, shall we say, an imperfect version of western democracy. But neither, for the most part, are they now totalitarian countries. Many of them have quite

a wide spread of traditional liberties. They are diverse in this. But in one characteristic they are similar: they have been economically successful.

(P. Hall, 1983a, p. 460)

As described earlier such an emphasis was also apparent in Howe's Isle of Dogs speech. He talked about the problems with democracy and advocated a kind of New Town Authority. As events turned out this aspect of his speech was taken up in the Urban Development Corporation initiative while EZs were to be administered by local authorities. But how much control do these local authorities and their local communities have over EZs?

It has already been noted that all the details of the planning scheme have to be agreed by the Secretary of State and therefore have to conform to the ideology of the initiative. It is also evident from Howe's speech in the Isle of Dogs that he is sceptical of civil servants and bureaucrats, which reflects the ideas of Kristol and others mentioned in earlier chapters. McAuslan (1981) has pointed out that one implication of this scepticism is a lack of consideration of the details of the policy and, as a result, the placing of considerable discretion in the hands of the Secretary of State. Details are thus worked out in negotiations rather than in public debate, inquiries or legal duties and powers. These negotiations take place between the DoE and the LA and between developers and LA officers with, in both cases, the former party holding the whip hand. McAuslan describes how both EZs and UDCs

> shift powers over land development from local authorities to central government and, equally important, shift discussion and debate about this land development from open forums, where the public could take part, to closed forums where negotiations between public authorities and developers will be the rule.

(McAuslan, 1981, p. 248)

The removal of the need to obtain planning permission also means the removal of the procedures and mechanisms to provide people with opportunities to state their views and protect their interests. As Purton and Douglas (1982) put it, 'by having a general perimeter definition of what constitutes permissible development, any proposal that conforms to it will by-pass the protective mechanisms covering individuals or bodies likely to be affected by the proposal' (1982, p. 417). The EZ procedure is a considerable streamlining of the normal development plan and development control procedures. The telescoping of the procedures in the EZ scheme has been variously

195

described as 'a hybrid of a local plan and a planning permission' (McAuslan, 1981, p. 250) and a 'unique mixture of local plan and planning permission' (Purton and Douglas, 1982, p. 416).

In the EZ proposal 'adequate publicity' has to be given to the scheme when it is devised. Objections can be made at this stage and the Secretary of State's designation order can be challenged within six weeks. There is no further opportunity to object and so there is no means of complaining about any proposal that conforms to the general criteria laid out in the scheme. There is also no requirement for the EZ scheme to conform with the structure or local plans which have been through a participation process. Compared to a local plan the opportunity to participate is significantly reduced.

The lack of public consultation is one of the consequences of the desire not only to remove planning interference in areas where it is considered unecessary but also to speed up any remaining proce-dures (see Keith Joseph in the *Financial Times*, March 28th, 1980). This desire has led to the removal of many decisions in EZ areas from the arena of the democratic local authority, for example, by changes to committee rules enabling many decisions to be taken by particular officers or councillors without the need for ratification (Tym, 1984).

Thus as far as the changes in procedure are concerned there have been significant implications by:

1 giving greater power to the Secretary of State;
2 giving greater bargaining power to developers;
3 removing opportunities to participate or object;
4 removing some decisions from local authority democratic debate;
5 diminishing the value of local plans;
6 telescoping plan-making and DC into one scheme.

What has happened in practice?

Having discussed the mechanisms devised for the simplified planning scheme of the EZs, what lessons can be learnt from their operation? Can the argument that planning regulation in the EZ schemes has not been abolished but simplified and re-oriented, be maintained (see Williams and Butler, 1982; Anderson, 1983)? An analysis of the exemptions and conditions of the EZ schemes shows that consider-able controls still operate and that there is much variation from zone to zone. In what direction has planning been simplified?

Much of the evidence for this evaluation comes from the monitor-ing reports of Roger Tym and Associates. However, the lessons that Tym could draw from their research were limited because 92 per cent of the floorspace built at the time of the survey had either received planning permission in the normal way, was developed by public agencies or was on land owned by public agencies and therefore subject to landlord control. Thus the effects of deregulation on design standards and land-use structure could not be truly assessed.

In the 8 per cent of floorspace agreed under the 'free' planning regime there was no obvious decline in design standards. Problems that were met related to access, landscaping and use of poor bricks on a prominent site. The reasons given by Tym for the good standards are;

1 the large degree of informal contact between developer and officers. 'Perhaps because they are not entirely accustomed yet to the freedom afforded by the planning scheme' (Tym, 1984, p. 121);
2 building regulations still apply;
3 standardisation of materials and construction;
4 buildings are investments and need to keep their value.

As indicated above, one of the implications of EZ schemes is the reduced control over design matters. Tym has shown how this control has largely been maintained in the EZ experience through using the power of landownership. Hence the control has shifted from the planning process to that of ownership and its supporting legal powers. The effect of the removal of design controls in other areas, such as those that might be included in SPZs where the ownership pattern is more complex and in private hands, is unclear. However, although planning controls over the detailed design of individual developments have been curtailed, some broader design aspects are still covered by planning. A prime objective of the EZs has been to attract development and hence the 'image' of the zone is very important. This has led to concern over the environmental appearance of the zone and the designation of sensitive sub-zones for special treatment.

In the case of *certain categories* of development planning permission can be assured from the outset. As indicated by the Tym report this allows for some speeding up of the process in the case of smaller applications. It also gives developers greater flexibility to modify the details of their schemes as they pursue the various stages of development preparation. It has been suggested that this reduces uncertainty for the developer and hence risk. However, this view has to be considered in the light of evidence showing that, in these EZ areas, there was little opposition to development in any case. Assumed planning permission is still subject to certain constraints covering 'reserved' matters, set out in the scheme, which are considered to have 'bad neighbour' or 'externality' effects. It was shown in Chapter Five how the exponents of less planning intervention do

not agree on how narrowly externalities should be defined. This uncertainty is carried through in the EZ scheme discussions. Initially
Howe took a very limited approach which was extended in the course
of the Parliamentary process. This was further extended in the negotiations over parking provision. The government was originally
against any controls over this issue but eventually conceded that if
developers did not provide parking provision it could lead to congestion and lack of safety on the roads and hence an externality problem
for other users.

What have been the effects of the planning schemes on the land-
use structure of the areas? In theory the zones could have been used
for office, residential, hotel, leisure and retail uses (within restrictions) as well as the industry and warehousing uses which were
usually dominant in the previous local plans. But in practice this has
not happened – retail uses in Swansea being the main exception.
Other smaller exceptions that have occurred are Salford and the Isle
of Dogs, where residential allocations in local plans have turned into
industrial use; Clydebank and Team Valley EZs, where offices have
been developed; a leisure centre in the Isle of Dogs and, as well as
retail, a hotel and conference centre proposal in Swansea. The Tym
report suggests that the land-use structure resulting from simplified
regimes in EZs is not that different from the pattern which would
have occurred under the previous system. However, the DoE seems
to contradict this in their summary of the EZ experience:

> It would not be true to say that all the development taking place
> in EZs would have been permitted by the authority anyway: there
> are some uses (for example, direct selling operations occupying
> industrial units) or aspects of development (for example, colour of
> buildings) that would not have been allowed previously. To that
> extent the planning freedom provided by the EZ has a real effect.
> (DoE, 1983c, p. 5)

Thus there seems to be a difference of opinion over this matter. However, the differences may be explained by a variation in the definition
of what constitutes a significant change in planning approach.

What are the implications of the changes to planning administration and how have developers reacted to these changes? One
important effect mentioned by the third Tym report is the removal of
protection for third party interests. However, in the EZs this has not
been a problem because the areas are predominantly industrial. The
worst effects, for example, noise and pollution, are covered by the
scheme anyway, and landlord control is operated by public authorities in a large number of cases. Tym also records how authorities
have changed their committee procedures to speed up decisions and

allocated a single officer, sometimes from the estates department, to act as a single point of contact with the developer. Hardly any developer regarded the simplified planning scheme as an important factor in their decision to develop, but they did like the greater speed and directness of contact. Some developers mentioned the certainty which the scheme provided, as it gave the requirements and conditions from the outset. It also allowed the developer the flexibility to chop and change in the detailed design process without having to go through further bureaucracy. Developers stressed the time-saving element of the scheme but Tym was not sure that this was really critical when compared with the time taken to sort out other aspects of the development process such as finance. Tym concludes that the major changes resulting from the new procedures were to increase the collaborative nature of the relationship between developers and the authority and to provide increased power for the developer in this collaboration. Thus the report concludes that 'liberation from bureaucracy has been less significant than the efforts made to achieve active collaboration between public and private sectors' (Tym, 1984, p. 151) and 'there can be no doubt that the negotiating position of developers is much stronger in EZs than it could ever be outside' (p. 138).

In the interest of quick decisions such methods as 'single person contact' and streamlined committee procedures have been adopted. The scheme also combines the normal processes of plan-making and development control. The net effect of this is to reduce public participation, third party involvement and opportunities for appeal. The shift of power to the Secretary of State and developer, both operating under a market-oriented ideology, means less opportunity for the planner and local authority decision-making mechanisms to use discretion in any particular case.

While the Roger Tym monitoring exercise was still under way the government announced a second round of zones. Designations were made in 1982 and most became operational in the following year. There is evidence to show that the objectives in this second round were somewhat different with an emphasis on the 'need to be effective quickly'. This led to the designation of green field and prepared sites where the state takes on less of an infrastructure co-ordination role (Lloyd, 1984b). Planning deregulation applies to a wider variety of areas, including outer city and rural areas, as a result of the new designations. This has ideological implications as deregulation is not restricted to run-down, derelict inner city locations. Some local authorities included in this second round, for example, in north-west Kent and Invergordon, wanted to prevent housing going into their EZs (see *Planning*, 1983a). They feared that the free operation of the market would lead to housing developments which would run counter to their policies but the government refused to allow housing restrictions. The greater variety of geographical areas and land uses in this second round provides a link to the subsequent Simplified Planning Zone (SPZ) proposal.

What kind of re-orientation of planning is suggested by these experiences of the EZ scheme? In referring to the 1980 Local Government and Planning Act of which the EZ initiative forms part, Loughlin (1981) has said that the 'shift is towards acceptance of the pre-eminence of the market mechanism and the consequent modification in the role of the State to that of a market aid rather than as an alternative allocative system to that of the market' (1981, p. 445). As far as the EZ planning scheme is concerned it could be said that there is a shift towards the dominance of the market in that life is made easier for developers regarding design matters and that the procedures have been streamlined for their benefit. As a result they gain power in the bargaining process. However, this has not necessarily translated itself into any impact on the land-use patterns on the ground. This could be because planning had already been following market trends in the EZ areas and second because controls were still applied to the important area of retail development. One of the shifts in the relationship between the market and the state that can be detected is the transfer from the private sector to the state of much of the cost of environmental improvement and infrastructure provision. This can be described as aiding the market by lowering developers' costs and hence reducing financial risk. Finally the state is being employed to remove matters from the public, democratic, domain to that of the market through the changes in the decision-making procedures.

What conclusions did the Tym research reach about the wider application of the EZ planning scheme? In their third monitoring report they were asked by the DoE, who had started to formulate proposals for SPZs, to look particularly at the possibility of extending the planning deregulation aspects of the EZs. The report states that the EZ scheme worked reasonably well and concludes that it would be feasible to extend the idea. However, it also warns that in areas of complex land-use patterns there are likely to be more 'third party' problems. The report assumes that SPZs will be used for difficult sites and infrastructure provision. The conclusions of the Tym report were not seen by the government as detrimental to the view already expressed within the DoE that 'EZ schemes and the planning regimes they contain would appear to offer some advantages over normal planning control for developers and for planning authorities seeking to encourage development' (DoE, 1983c, p. 7). The green light was therefore given to the announcement of the SPZs.

Extending the approach: Simplified Planning Zones

The wider application of the simplified planning regime was part of Howe's thinking when he first launched the EZ idea. It is no surprise therefore to find that the regime was extended into a 'second round' and later into SPZs regardless of any findings of the research into the experience of the first round. The extension of the EZ planning scheme had been under consideration in the DoE long before the publication of the Tym third monitoring report. For example a paper was produced in June 1983 to report on this matter to ministers and it contained some draft ideas on extending the EZ planning scheme (at this time referred to as Local Development Scheme). The introduction to this paper contains some interesting comments on the purpose of extending the simplified planning regimes:

> Ministers want us to consider whether the 'simplified planning regime' of EZs could be extended more widely. Clearly one would not want to use the Parliamentary process required for EZs, which is necessary because of the major fiscal implications of EZs. But the principle of defining an area within which the limits of planning control (and other essential requirements) are defined, and removing conventional discretionary planning control over other types of development, offers a very interesting alternative to the system that has existed since 1947 – and which some may consider to be increasingly anachronistic. The area approach offers the possibility both of testing the simplified system and also perhaps of changing attitudes towards the proper purpose and extent of development control.
>
> (DoE, 1983c, preface)

This statement indicates clearly the role that an extension of the EZ planning scheme could play in reformulating the whole framework of the planning system. This would seek to replace the 1947 system and also change the whole purpose and extent of development control. These sentiments contradict the view that EZs and SPZs are only tinkering at the margins of the planning system. The other interesting aspect of the statement is its clear exposition of the ideological role played by these policy initiatives in 'changing attitudes' to the 'proper' role of planning.

SPZs were first publicly mentioned in Jenkin's speech to the RTPI Summer School in September 1983 (Jenkin, 1984). A consultation paper, based on the above DoE draft, was produced in May 1984. The paper suggested that when a local authority initiated an SPZ it should consult government departments and other public and private agencies to get their agreement and consider objections either through written statement or public inquiry. At this stage the paper did not specify the exact nature of the procedures but set out different alternatives. Examples of the kinds of areas where SPZs might be used were stated as:

1　in commercial areas, giving more flexibility to possible changes of use between offices, light industry, science parks, etc.;
2　in residential areas, leaving details of layout, design and landscaping to the developer;
3　in areas of urban regeneration, allowing a more promotional approach.

The White Paper *Lifting the Burden* (DoE, 1985a) included the SPZ as a major example of planning deregulation and set it within the broader philosophy of deregulation. The paper contains a general attack on planning which is blamed for causing delay, uncertainty and using discretionary powers to impose excessive detail. As far as SPZs are concerned the White Paper says that the zones could be a 'stimulus to the redevelopment of derelict or unused land and buildings' (1985a, p. 11) and significantly, that SPZs can be initiated by private developers or the Secretary of State.

Comments on the consultation paper and *Lifting the Burden* stressed a number of issues (see for example, GLC, 1985; Lambeth London Borough, 1985; TCPA, 1984; Edwards *et al.*, 1986). The point was made that research evidence does not demonstrate the negative effect of planning regulations. Of particular concern was the increased centralisation of control and reduction of participation and right of objection which the London Borough of Lambeth feared could allow the Secretary of State to overturn council policy. Doubts were also expressed that competition between authorities could stimulate the unsatisfactory lowering of standards in SPZs.

Notwithstanding such criticisms the government pressed on with the idea and included the necessary legislation in the 1986 Housing and Planning Act. In November 1987 the SPZ Regulations came into force and guidance was provided in Circular 25/87 (DoE, 1987d), elaborated in a Planning Policy Guidance note in 1988 (DoE, 1988c). In introducing the concept of the SPZ this note shows how the scheme creates an alternative system to the normal planning controls:

> SPZs provide an alternative approach to the control of development which will allow the planning system to work more efficiently and effectively in the designated areas and thereby encourage development to take place were it is needed.
>
> (DoE, 1988c, para, 3)

This approach is seen as allowing local authorities to stimulate development through its designations in the SPZ scheme, and to generating confidence in the area covered by the scheme. However, as Lloyd (1987) points out, this creates a two-tier planning system and creates potential uncertainty and a more complex system (see also Stungo, 1985).

The DoE guidance note also makes it clear that the purpose of the simplified regime, as indicated in *Lifting the Burden*, is to make it easier for developers to respond to market forces. In exhorting local authorities to keep conditions in SPZ schemes to a minimum the note says; 'the greater the degree of freedom conveyed by an SPZ scheme, the easier it will be for developers and landowners to respond to client preferences and market conditions' (DoE, 1988c, Annex, para. 12).

The procedure eventually chosen by the government for the adoption of an SPZ is that laid out for the adoption of a local plan. This means a fairly lengthy and complicated procedure involving consultation, placing the scheme on deposit, allowing for objections and holding a public local inquiry. This would therefore allow for participation although the periods allocated are short and, as with local plans, the emphasis is on the formal mode of participation. However, once the scheme has been adopted there is no further right of objection to specific development proposals that conform to the scheme. Thus participation can be only at the level of general principles and a person cannot later object because of the local detrimental effects of a particular development proposal. This approach does not allow for the flexible treatment of development proposals according to local criteria. Once the scheme is agreed it can be modified only by a lengthy revision procedure, which reduces flexibility and could make it difficult to respond quickly to changes in market conditions. This is a criticism often made of the zoning system in the USA on which the SPZs seem to be based (Lloyd, 1987).

The procedures give the Secretary of State considerable reserve powers. If it is considered that the local authority proposals are unsatisfactory, the Secretary of State can order modifications or can direct that the proposals are submitted to him/her for approval. The Secretary of State also has default powers to prepare a scheme if a local authority appears unwilling to do so. Of particular interest is the facility which allows developers to request a local planning authority to make or alter an SPZ. As the guidance note says, in such circumstances 'it is hoped that authorities will respond constructively to proposals for SPZs put to them by developers or land owners' (DoE, 1988c, para. 8). If they do not then the Secretary of State can direct them to do so. This is an interesting shift of initiative in the control of the planning system. However, it is not clear whether this facility will generate much enthusiasm. Roger Humber, the Director of the House Builders Federation, is very sceptical of the proposal and finds it 'hard to see precisely what Ministers expect to achieve by this remarkable proposal which, in effect, invites local authorities to commit *hara-kiri*' (Humber, 1986, p. 3). He suggests that the government is trying to 'implicate private developers in cross-party political battles' and that the government should introduce the schemes themselves and not expect others to do it for them.

The kinds of areas suggested for SPZ treatment in the final version are similar to those indicated in the consultation document but with perhaps more emphasis on the regeneration of older urban areas. Specific kinds of areas that are mentioned are new industrial parks, old industrial areas, large disused sites, large single-ownership sites, and large new residential areas. As indicated in the earlier statements the SPZ planning scheme follows very closely that of the EZs, setting out the kinds of uses that can be regarded as having deemed planning permission. Two alternative approaches are suggested: the specific scheme or the general scheme. In the first approach the types of permitted development, and any limits on these, are set out. Any other proposals will be subject to the normal planning system. In the general scheme the starting-point is to allow any kind of development and then list any exclusions. As with EZ planning schemes, certain conditions and limitations can be imposed although these are supposed to be kept to a minimum. Such conditions are to be set out in clear development criteria and can cover health and safety standards, form and scale of development, heights and density, floorspace limits and parking standards. Landscaping and highway safeguarding sub-zones can be designated. As with many of the other deregulation initiatives, environmentally sensitive areas such as National Parks, conservation areas and Green Belts are excluded from the legislation.

Conclusions

Can the move towards simplified planning regimes be said to reflect the principles of Thatcherism? Can one detect the influence of the neo-market strand in changing the basic criteria on which decisions are based? The original speech by Howe on EZs, *Lifting the Burden*, and the recent Planning Policy Guidance Note on SPZs all see the new regimes allowing greater freedom to the market. The attacks on planning by the Thatcher governments and the undermining of the post-war consensus on intervening in 'the public interest' allow developers to push their case with greater confidence and alter the balance in the power relations between developers and the local authority. There is evidence, for example, in the Tym report, that this has happened in the EZ planning schemes. However, EZs have generally occurred in areas suffering from the lack of investment where local authorities take a flexible attitude. It is in the extension of the idea to SPZs that greater controversy could arise. The facility for developers and the Secretary of State to instigate SPZs is significant in this respect. If developers or the Secretary of State employ their power in determining the location, extent and content of these zones, then this can be regarded as a significant change in the process of planning and would further limit the likelihood of non-market criteria being adopted.

However, the resulting schemes are not as 'planning free' as Sir Geoffrey Howe and others had hoped when the simplified regimes were first considered. In the EZs planners still control certain market externalities and sensitive areas. This role is extended to SPZs. One question here is how restrictively 'externalities' and 'sensitivity' are defined. The clear message in the government guidance is that such controls have to be kept to a minimum and if they are not the Secretary of State will use his/her call-in powers. In addition planners have intervened in the market with respect to retail use. This could be seen as an example of planners retaining a role where two economic interests, small and large retail interests, are in competition and where one manages to secure the ear of government in protecting its particular interests against open competition. This would appear to be a dilution of the ideological principle, although *Lifting the Burden* suggests that ideological purity should be restored. The failure of the attempt by certain authorities to extend the intervention to a restriction on housing development indicates the tight limits on any relaxation of the market principle. Thus the state continues to play some kind of regulatory role over the market. It has also been seen that the state plays a subsidising role in providing finance, for example, through rebates and infrastructure provision. This takes some of the burden from developers and aids the functioning of the market in high-risk areas.

The procedures for EZ planning schemes reduce the opportunities for participation and give increased power to central government. Negotiation in private meetings is the basic mode of decision-making. In the extension to SPZs the process is more public through the adoption of the local plan procedure. However, the Secretary of State still has considerable reserve powers. The adoption of simplified planning regimes alters the basic processes of the planning system in the areas in which it is applied regardless of the more dramatic provision of allowing developers and land-owners to instigate an SPZ. The result is the co-existence of two different planning processes. The basic difference between the two procedural approaches is the omission in the simplified regimes of case-by-case decision-making and the associated administrative discretion. The replacement of this discretion by a general schema could be seen as a major change in the whole approach to the control of development. The new approach could be said to be a move towards Hayek's 'rule of law', setting out a framework that is known and agreed beforehand and within which subsequent decisions must conform. It has also been suggested (for example, Lloyd, 1987) that the simplified regime approach could lead to greater reliance on legal means of solving bad neighbour problems. *Lifting the Burden* indicated that the laws of nuisance were under review reflecting the suggestions in the Omega Report (Adam Smith Institute, 1983). The Tym report shows that in the EZs most planning control has been exercised through the means of landownership. As a result enforcement and resolution of any disputes are moved from planning to the legal arena. The lack of flexibility and opportunities for objection at the detailed level that result from this alternative approach have already been mentioned. Any opportunities arising from 'planning gain' are also lost. Although planning gain may not allow for participation it does create opportunities for using non-market criteria to obtain community benefits from development.

One of the implications of this withdrawal of administrative discretion is that the criteria that can be applied in the decision-making process are likely to be severely constrained. In the simplified regimes there is less public participation and less control by the local authority resulting in less opportunity to press for non-market criteria in planning decisions. *Lifting the Burden* talks about the need to maintain a balance between the protection of the environment and fostering a prosperous economy. It is implying that not only do we need less planning but also that the planning that remains should be restricted to environmental protection. No planning contribution towards satisfying community needs is envisaged here.

The development of the simplified planning regimes can be seen to have passed through a number of stages. When the idea was first advocated by Hall, and taken up by Howe in 1978, it played a significant ideological role in the attack on regulation. Such a role has been maintained through the statements of pressure groups such as the Adam Smith Institute. Meanwhile the idea has been translated into action with various concessions that have somewhat diluted the ideological purity. This implementation was first expressed as an experiment with the hope of extending the initiative to other areas later. This has happened with further EZs and then the SPZs. There has been a shift from an 'experimental' attitude to one of considering the simplified regimes as an alternative to the 'normal' planning system. As a result there is now a dual system operating.

Chapter ten

Conclusions: the re-orientation of the planning system

This book has explored the interface between political ideology and the planning system. A major question has been whether the legislative changes since 1979 are simply small adaptations or add up to an important shift of direction. A detailed account of all the different changes, often small in themselves, has been presented in order to investigate possible patterns. A key purpose has been to search for coherence or consistency. The book has examined not only the modifications to the normal planning system but also the alternative approaches that have been adopted. This concluding chapter explores how far this has led to a re-orientation of the planning system and discusses some of the implications for the future.

Principles of decision-making

Economic liberalism is an essential strand of Thatcherism. According to this philosophy only the market can deal with complexity and encourage innovation and hence all decisions should be based upon market principles. The use of any other decision-making criteria is to be avoided. Inequality is seen as necessary and positive. Hayek, Friedman and Joseph attack the meaninglessness of concepts such as 'social justice', 'fairness' and 'equality' and, as the post-war Welfare State is built around such concepts, its value system is subjected to attack. The role of the state is re-oriented to support rather than supplant the market and this support can take the form of providing a framework of infrastructure, or legal and financial provisions. The state can also make the market more efficient through dealing with the problems of 'externalities' or 'neighbourhood effects'. It is also accepted that there may be people who cannot obtain their needs through the market and so require state assistance. The exact extent of 'externalities' and people needing help is not clear; however, it is considered that both categories of state action should be kept to an absolute minimum. The New Right authors on planning adopt these principles and advocate that land-use and development decisions should also be left to the market as far as possible. Cities are seen as too complex ever to be understood by planners whereas the market can deal with this complexity. Planning is also a cost burden to developers and the country as a whole and detrimental to wealth creation. These writers also attack the meaninglessness of the non-market criteria adopted by planners, such as 'public interest', 'neighbourhood', or 'self-containment'. However, it is accepted that planning can fulfil certain roles that would support the market such as dealing with externalities, conservation, providing future-related information or infrastructure. However, again there is no agreement over the exact extent and boundaries of these activities.

How far are these principles of decision-making evident in the changes to the planning system since 1979? There is no doubt that underlying the initiatives is the desire to 'free-up' the planning system thereby giving greater scope for developers and housebuilders. This market freedom is the essential ethic of the Urban Development Corporations, the Enterprise Zones and the Simplified Planning Zones. Many of the modifications to the planning system remove constraints on developers, for example, the relaxation of controls on industry and small business or the changes to the Use Classes Order. Part of the process of giving greater freedom to the market has involved the downgrading of alternative decision-making tools such as plans and policies. Such statements have been reduced in significance as they have become only one 'material consideration' alongside market pressures and demands. In line with the general attack on alternative criteria there has been the reduction in the acceptable scope of planning intervention. Thus design controls and social criteria have been much reduced or even excluded, for example, in the modifications to structure plans or the appropriate scope of planning conditions or planning gain agreements. The UDC experience has shown the reliance on spin-off effects to deal with social considerations. It is argued that if the market is allowed to flourish to its maximum extent without hindrance then the benefits of greater wealth generation will filter down to everyone's advantage. This implies that no positive or direct action to meet social criteria is needed.

Although there is clear evidence that the changes have restricted the scope of planning and reduced the permissible criteria for decision-making, it is also clear that certain functions of the planning system have been retained. The changes have not abolished the system itself but restricted its scope. Functions still remain as demonstrated in the development of the EZ initiative, where control was retained over externalities, sensitive environmental areas and infrastructure. The exact extent of these roles, both in simplified planning regimes and other planning control mechanisms, remains open to negotiation which reflects the uncertainty in the theoretical literature. A further area of uncertainty is evident in the debates over controls on retail use in the simplified zones. This demonstrated that sometimes a conflict can arise between the general ideology of the market and lobbies pressing for the protection of a particular economic interest.

Procedures of decision-making

'Authoritarian decentralism' has been identified as another aspect of Thatcherism. The term has been used to convey how decision-making power has been centralised and then redeployed to the market-place. This process involves a downgrading of the importance of democracy and participation. Authors such as Friedman and Hayek criticise political processes because they are subject to pressures and influence, involving corruption and bias. They extol the virtues of the market as an alternative. This theme is taken up by some of the New Right writers reviewed in Chapter Five who decry the effects of public involvement and the politicisation of the planning process and advocate that land use and development should be left to the experts, the developers themselves, who take the risks and have the requisite knowledge. Hence the proposal that planning decisions should be removed from their current setting, where they are exposed to political and community pressures, and placed in the more 'abstract' arenas of the law and market.

How far have these procedural ideas been reflected in practice? First there is a very clear centralisation trend running through all the changes. For example, this is evident in the more interventionist role of the Secretary of State, the increase in appeals, the use of Circulars and the introduction of more reserve powers in the legislation. As mentioned above, central government has been keen to restrict the scope of planning and the range of criteria used by local authorities. A very direct increase in central control is introduced throughout the initiative of strategic guidance. This potentially allows central government to set the parameters for all subsequent plan-making. The Urban Development Corporation initiative is another clear example of the 'authoritarian decentralism' principle. Central government has very strong controls over planning in these areas through its financial provision and the role of the Secretary of State in setting up the Boards and arbitrating between any conflicts. This centralised power is then used to ensure that decisions are made according to market priorities. Simplified Planning Zones also demonstrate the same principle; for example, these zones can be imposed upon local authorities by the Secretary of State and developers can also request such designations. In such zones it will not be possible to pursue planning gain and so one of the remaining mechanisms for local authorities to introduce community-oriented criteria will be lost.

The reduction of participation opportunities is another thread running through the changes. As the RICS (1986) points out, public involvement has the effect of broadening the criteria included in the discussion and so an antagonism to participation is a logical part of the New Right perspective. The speeding up and streamlining of the planning process has led to less participation. Many of the new approaches such as the planning procedures in Enterprise Zones and Urban Development Corporations involve reduced participation. In the UDCs the Code of Consultation was seen as the mechanism for local involvement but experience has shown that this has not been given any priority. The effect of the procedures in Urban Development Areas is to by-pass local democracy and community involvement and redefine democracy as purely based on the general election. Under this broad legitimacy decisions are taken in the private arenas of the market and central government departments. Also design considerations are thought to be inappropriate for democratic discussion and should be left to the experts who advise developers.

Anti-bureaucratic sentiments

A third key element of Thatcherism has been termed 'anti-bureaucratic sentiments'. Thatcherist ideology incorporates a dislike of bureaucrats because of their basic antagonism to capitalism and the market. This antagonism is seen as built upon self-interest and bureaucrats are seen to justify their role through the use of such concepts as 'the public interest'. According to Thatcherism their power needs to be kept in check and their views disregarded. This attitude is also evident in the New Right approach to planning and the values of professional planners are attacked, for example, for the way they under-mine property rights. The Omega Report refers to a 'planning class' with its own interests and attitudes which are detrimental to society.

One dimension of this attitude towards bureaucrats is the dislike of administrative discretion. Not only does this allow bureaucrats to impose their own values but also it provides the opportunity to employ the non-market criteria that are considered unacceptable. Hayek's 'rule of law', in which discretion is replaced by a known set of generalised and impartial rules, can be seen as one response to this. These rules would be supportive of the market and property rights and in particular treat everyone on an equal basis preventing political intervention to instigate positive discrimination. Such a set of rules would be administered outside the political arena and government would be obliged to conform to them. Support for this approach is evident in some of the New Right ideas on planning where preconceived frameworks and legal processes, such as covenants and laws of nuisance, are advocated. However, politicians pursuing the Thatcherist ideology have not taken up such ideas as the 'rule of law' with any enthusiasm. In fact the authoritarian strand of Thatcherism and the processes of centralisation of government power illustrate Hayek's point that a Conservative government has a tendency to want to ensure and direct orderly control. This creates tension with the liberal notion of limited government and Hayek's wish that the 'rule of law' should provide a restriction on the power of government.

The anti-bureaucratic sentiments dominate the government's attitude to planners. This is illustrated by the single-minded way in which the government has pursued its ideas without discussion, and often in the face of opposition from the planning profession. The centralisation process which runs through the period has the effect of reducing local bureaucratic discretion, although this is usually replaced by central government ministerial discretion. Simplified planning regimes in particular directly remove case-by-case treatment. There is some evidence of moves towards the kind of general preconceived frameworks mentioned above, for example, the area frameworks of the UDCs, the simplified regimes of the EZs and SPZs. However, there is no evidence of a desire to move towards covenants or a universal zoning system.

A coherent attack on planning?

The economic liberal and authoritarian strands of Thatcherism have clearly affected the planning system through the strengthening of market criteria, increased centralisation and downgrading of public participation and bureaucratic discretion. However, potential contradictions in the ideology have also been detected and these have influenced the impact of the ideology on planning. To what extent can the changes to the planning system be said to have a coherent direction?

One potential contradiction relates to the purpose of planning. The scope of planning has been much reduced and oriented towards the market. Thus economic promotion and the reduction of constraints on economic interests, such as small businesses and offices, have been paramount. However, there are many occasions when this market supportive stance conflicts with environmental protection and amenity. Indeed the governments since 1979 have continued to express the value of conservation and protection of the environment. This economic/environment conflict is evident throughout the planning system. Examples discussed earlier in the book include the difficulties in operating the development control circulars that relax regulations, the introduction of 'sensitive areas' in simplified planning regimes and conflicts in local plans, as exemplified in the debates over the draft City of London Local Plan.

The Green Belt is another example of the conflict between economic development and environmental protection. This crystallised in the saga over the 1983 and 1984 Circulars. Elson (1986) provides an excellent description of the divisions within the Conservative Party that were exposed in the battle over these Circulars. The lines were drawn between, on the one side, the housebuilders and those elements in the party representing property interests and on the other the Council for the Protection of Rural England and politicians supporting their constituents who were concerned about the preservation of their particular piece of the countryside. According to Elson the battle can be seen as representing the wider issue of the government's commitment to land-use planning as a legitimate public sector activity. He shows how, in a period of recession, the Green Belt is vulnerable to pressures from housing, mineral and retail economic interests. These interests can lobby for the relaxation of controls in order to promote economic progress. In the end the 1983 Circular, which sought to shift the balance between conservation and development towards the latter, was replaced by the 1984 Circular, which reflected contemporary practice. The Green Belt example shows the potential of environmental protection interests to check the progress of market-oriented deregulation.

The conflicts over economic and environmental criteria are also well illustrated by the controversy over the proposed new housing settlements in the South East. This controversy also shows how the lack of a planning framework and the relaxation of regulations can create problems within Conservative ranks. Central government, responding to pressure from the housebuilding lobby, has been inclined throughout most of the last decade to consider favourably new settlements in greenfield sites. This again creates antagonism from local Conservative supporters keen to preserve the environment and their own residential amenity. The extent of this antagonism was illustrated in the huge increase in the Green Party vote in the 1989 European Parliament elections. The Party attracted 21 per cent of the vote in the Thames Valley and about 20 per cent in Hampshire, both areas under particular pressure from these new settlements.

The approach of the developers in trying to get permission for these settlements also illustrates the procedural trends described throughout this book. The developers have scant regard for local policies and assume that they will go to appeal and present their case at an inquiry. This means that the decision will be taken by the Secretary of State, who has been generally seen as sympathetic to their cause and who will often override the advice of the inspector. Here again the centralisation process can be seen at work. Whatever the outcome central government has been directly involved in development decisions.

This conflict between market criteria and environmental protection can be seen to reflect cleavages within the Conservative tradition. The environmental protection lobby is often supported by the land-owning element in the Conservative Party who, as discussed in Chapter Three, prefer the ideology of One Nation paternalism. The differences are well illustrated by the exchange of letters in March 1988 between the Secretary of State for the Environment, Ridley, and the former Secretary, Heseltine (*The Planner*, 1988a). Since leaving the front benches Heseltine has frequently voiced the opinions of the protectionist faction. In his open letter to Ridley he expresses his concern that the countryside in the South East is being ravaged by the pace of development and he asks Ridley 'to throw the weight of his Department on the side of environmental restraint'. In his sharply worded reply Ridley claims to be applying considerable restraint but goes on to say that 'it is easy to point at the "greedy" developers. But they only exist because they have customers. Housing is not a form of environmental pollution. It is about people and families, where they work and where they live'. The two continued their dialogue in the speeches they made to the Conservative Party conference later in the year (*The Planner*, 1988b). Ridley focused on the improvement to the quality of new housing as the solution to conflicts with the environment. Meanwhile Heseltine's approach was one of regional development and inner city investment to relieve the pressure on greenfield locations in the South.

As already mentioned, one of the ways the government has reacted to the dilemma created by this conflict over the appropriate purpose of planning has been to abandon the principle of a universal planning system. The changes that have been made to the legislation since 1979 have accentuated the differences between areas. As a result there is a dual system operating. There are those areas with high environmental value where strict controls still exist and in which administrative discretion still operates. These areas are the National Parks, Areas of Outstanding Natural Beauty, Conservation Areas and Green Belts, all of which have been exempted from the relaxations to the planning legislation. Then there is the much more relaxed 'normal' system in which there has been a shift to a greater acceptance of market criteria and a downgrading of development plans and policies. Thus the response to the economic/environment conflict has been this disaggregation of the planning system.

It could be said that this acceptance of environmental criteria is a watering down of the ideological purity of economic liberalism. One way the relaxation of ideological purity might be justified is through the concept of 'externalities'. Under Thatcherism it is acceptable for planning to minimise the effects of externalities and ensure the maximum efficiency of the market system. It might be said that the protection of the environment falls within this acceptable role. In this way the relaxation of ideological purity could be seen as merely a minor adjustment and the dominant force of economic liberalism retained without the need to raise questions of alternative principles or values. The problem is how far such a concept as 'externalities' can be stretched and, as already noted, the academic writing is very unclear on this.

Another potential contradiction has already been mentioned. This is the tension between the desire for a 'rule of law' which provides a generalised framework which stands above government and, on the other hand, the tendency of government since 1979 to centralise power. The first principle encompasses an attack on individual discretion while the second tendency gives considerable discretion to ministers. It has been seen that many New Right academics, in particular Hayek, support the idea of the 'rule of law'. As far as planning practice is concerned it has been noted that certain initiatives could be interpreted as moving in this direction, for example, SPZs, EZs, and area frameworks. There are also strong lobbies on government, for example, the Adam Smith Institute and British Property Federation, who wish to see these kinds of approaches extended. However, as implemented so far these approaches do not reflect all the requirements of the 'rule of law' idea. Although they contain preconceived frameworks and remove local case-by-case discretion they shift control to central government. Thus there has been a considerable erosion of administrative discretion as previously conducted at the local authority level but central government since 1979 has retained its power to oversee and orchestrate changes in the system. The Thatcher governments seem reluctant to give complete responsibility to the processes of the market or the legal system.

The reluctance to push such approaches as simplified regimes, zoning or restricted covenants any further could be interpreted as a tension between two strands of the ideology. Approaches based on the principles behind the 'rule of law' conform to the economic liberal aspects of the ideology. However, the authoritarian strand supports and seeks strong government authority. In practice the result has been a strong interventionist central state which has sought not only to create a change in the values and criteria behind decision-making but also to direct the shift in operations to the market-place. This approach was labelled earlier 'authoritarian decentralism'. Authoritarian values have been employed to undermine the processes of democracy as it operates at the local level, and thus community involvement and the power of local councillors have both been eroded. The impact on the planning system has been significant as local authorities have been the arena in which most planning decisions have been taken in the past.

This downgrading of local democracy and local administrative discretion has also led to a further disaggregation of the planning system. It has already been noted how conflicts between environmental and economic criteria have led to a differentiation in the system. That part of the system which gives priority to economic criteria has been further disaggregated, this time in terms of the different procedures adopted. In one part of the system there are procedures which, although much modified and downgraded, are still basically grounded in the 'normal' local democratic arena. In the other part of the system concessions have been made to some of the principles underlying the 'rule of law'. Here local democracy is avoided or distorted, participation and administrative discretion minimised and preconceived frameworks favoured. Such a procedural approach has not been applied in any pure form and often initiatives contain elements of both approaches. However, a distinct and alternative pattern to normal democratic processes can be detected in the procedural approaches of initiatives such as simplified planning regimes and UDCs.

These tensions reflect the debates over the very nature of Thatcherism discussed in Chapter Three where inherent contradictions between the economic liberal and authoritarian strands were identified. As discussed in that chapter, views differ over whether these contradictions make the whole ideological edifice precarious and vulnerable to collapse or whether the resolution of these contradictions leads to the uniqueness of Thatcherism, giving it strength and flexibility. Such issues are clearly evident in this study of the planning system. It has meant that at times the purity of the ideology has been tainted yet the questioning and erosion of planning has continued. This is illustrated by the history of the simplified regimes in the EZs. These were modified from Howe's original conception but the idea was extended in the SPZ even though research did not back up the claims made regarding the advantages of deregulating planning.

Another way in which the ideology can have a momentum beyond its application to a particular policy initiative is through the permeation of attitudes to other aspects of the system. Thus the greater freedom available to developers in UDCs or EZs can create a climate in which challenges are made in other circumstances, for example, opposing a local plan or taking an application to appeal.

Another issue raised in the discussion over the nature of Thatcherism was whether any theoretical cohesion fell apart under the pragmatic pressures of practice. This study of planning shows that the ideological principles have been strongly reflected in practice and that certain means have been found to cope with potential contradictions within the ideology. Thus the geographical differentiation in the implementation of the changes and the application of many initiatives to specific areas has avoided exposing the contradictions. This would support the idea that the co-existence of economic liberalism and authoritarianism has provided the scope to fulfil a radical yet flexible programme with the ability to manage conflict within the ideology.

How radical a change?

Chapter Four analysed how the ideology of Thatcherism has led to a re-orientation of the role of the state. In general terms the state has been 'rolled back' in some areas, utilising the principles of economic liberalism, but strengthened in other areas such as law and order, employing the authoritarian strand. The conclusions regarding the changes to planning also indicate a re-orientation. This has involved a loosening of state regulation in some areas but, on the other hand, increased centralisation of the system. The changes to planning can be summarised as a re-orientation of the *purpose* of planning towards greater acceptance of market criteria, selective application of environmental criteria and the removal of social criteria. There has also been a re-orientation of the *procedures* of planning away from community-based local democracy towards centralised government supervision. In order to implement this re-orientation without exposing contradictions the principle of a universal planning system has been abolished and a disaggregated system developed. This disaggregated system is comprised of a division based upon the relative importance of economic and environmental criteria and a further division based upon the acceptance or not of local democratic processes. The country can therefore be divided into three areas operating different kinds of planning systems: areas in which environment is important and planning controls are strong; areas where economic criteria dominate but decisions are still made within the much modified and constrained procedures of local democracy; and lastly, areas where economic criteria also dominate but where decisions are made outside the framework of local democracy.

But how radical are these changes? The conclusion of the historical review in Chapter Two was that ever since the erosion of the 1947 Act's financial provisions, the powers of planning have been limited, that the initiative has always been with the private sector, and that property rights have always dominated. To the extent that these attributes have continued, nothing has changed. It is sometimes argued that the only difference since 1979 is that of perception – reality rolls on unaffected. In other words market domination has always been the case but under the post-war consensus this was hidden by the myths of welfare ideology. Thus it is claimed planners may have sought to pursue social objectives in the past but their effects have always supported the private sector (see Reade, 1987; P. Hall *et al.*, 1973). This view that planning under Thatcherism is no different from planning during the post-war period is rejected. The detailed study of Thatcherist ideology and its influence on planning legislation points to a significant re-orientation of the planning system.

The establishment of the 'Welfare State' and the corporatist approach to government were based upon the notion of a co-operative partnership between economic interests, the government and the community. This approach helped the state to both promote economic growth, with the resultant profits for individual economic interests, and at the same time maintain social stability. This was an approach that Conservatives also pursued based upon the 'One Nation' philosophy. As we have seen this whole post-war approach has been attacked and undermined by Thatcherism. Although in the post-war period the powers of planning were limited there was a legitimate basis for pursuing social objectives or community needs. The acceptance of intervention and the attempt by the state to respond to the needs of the population meant that concepts such as 'public interest', 'positive discrimination' or 'redistribution of wealth', could be employed to pursue policies and programmes that were not simply following market criteria. Thus although economic interests such as property developers, might always have had the upper hand there was also the scope and legitimacy to employ these alternative principles to gain particular results. Since 1979, with the arrival of an ideology that seeks to make market criteria even more dominant this opportunity has been lost. The chance of employing socially oriented concepts has been whittled away in the legislative changes over the period. The only significant opportunity that now remains to employ such community-based criteria lies in the operation of 'planning gain' and even this method has the disadvantage of limited democratic involvement. As a result of these changes it is argued that the scope and purpose of planning has undergone a major shift since 1979. During the post-war period planning was fulfilling three different purposes, though often in a confused or veiled fashion. These purposes covered the promotion of economic efficiency, the protection of the environment and the fulfilment of community needs. Since 1979 the first of these has become paramount, the second important only in specified geographical areas and the third no longer seen as the remit of planning.

The shift is reinforced by the authoritarian strand of the ideology. During the 1970s there were increasing demands for a more responsive and involved approach by government. The elitist and centralist nature of post-war government, including the provision of services by the Welfare State, was under question (Gyford, 1985; Boddy and Fudge, 1984). Since 1979, the authoritarian approach with its disregard for democracy and public involvement has brought an abrupt end to this movement. One result is less influence over development decisions. An example of this is the Channel Tunnel where special Parliamentary procedures for deciding the routes have been adopted. This is another clear case of the centralisation of the planning process with diminished opportunities for participation and consultation. The removal of public participation, the removal of local democratic influence, the anti-bureaucratic sentiments and the scorn of professional planning opinion have all reinforced the centralisation process and the shift in purpose away from community needs and aspirations towards the imperatives of the market-place.

Certain elements of the planning system are surviving; local plans have been retained, strategic thinking is continuing through strategic guidance and alternatives such as zoning have not taken a hold. However, it is not simply the outward form of planning that is important. It is absolutely crucial who has control. The trends identified in this book show a clear process of centralisation and a narrowing of legitimate criteria and values. A major change in the control and purpose of planning has taken place even though certain policy instruments remain.

The dominance of Thatcherism

Although the ideology of Thatcherism has been implemented at different speeds in different areas of social policy there is evidence of a general trend towards radical change (Flynn, 1989). This indicates that the re-orientation of planning identified in this book is just one reflection of the all-pervasive influence of Thatcherism.

Some policy areas have a close relationship to planning and so the impact of Thatcherism in these areas could spill over and affect planning. Two such areas are housing and the financing and organisation of local government. Major changes have occurred in these areas since 1979, such as the sale of council houses, the lack of housing finance, the reorganisation of housing management, massive reduction in local government finance, the ability of local authorities to control and determine their spending, the abolition of Metropolitan Counties and the lack of any regional policy (for a good analysis of these aspects see Duncan and Goodwin, 1988).

The same messages will be found: market dominance, and centralisation accompanied by downgrading of local political influence and public participation. The result is widening inequalities and a neglect of needs not expressed in the market-place. As more and more activities are taken out of state control and placed in the market-place so it becomes more difficult for planning to play any land-use balancing or co-ordination role. This also applies to infrastructure, the planning of which is often considered a legitimate role even from the right-wing perspective.

This book has concentrated on the analysis of ideology and its effects on planning legislation. The interaction of ideology and legislation has created a framework which constrains the operation of planning practice. However, this is not a one-way process and ideology cannot be viewed in isolation. Thus a more complete picture would be achieved by relating the conclusions of this study to those that examine the operation and implementation of planning. These latter studies would introduce a greater emphasis on power relations and the interplay of different economic interests and other groups in society. It would be interesting to build up an analysis which incorporated economic change, the influence of ideology, legislative change, mediation of interests and actual implementation.

In their study of a number of case studies Brindley *et al.*, (1989) identify a wide variety of what they call planning 'styles'. However, they believe that this variety and experimentation is drawing to a close and that planning practice in the 1990s will be dominated by the market-led approach. A thorough and wide-ranging analysis of the factors that have created variety in planning practice from the mid-1970s to the mid-1980s is provided by Healey *et al.* (1989). They conclude that the operation of planning has benefited certain interests but that conflicts between these interests is also evident. According to the authors these conflicts provide opportunities for the access of less dominant groups and the application of a range of criteria. It has been the contention in this book that the legislative changes over the last decade have greatly constrained these opportunities. It would be interesting to explore how much impact this constraint has on practice and to explore further the way in which ideologically inspired changes to the legislation have generated a shift in attitudes on the part of both developers and planners.

The experience of Thatcherism shows that planning can be significantly affected by ideology and that the definition of its role is essentially a political decision. This was also the conclusion of the Nuffield Foundation's review of planning (1986). There can be different views over the 'proper' role of planning. This means that it is no longer possible to reach a consensus on planning based upon a vague notion that it is a 'good thing'. Therefore any speculation over the future of planning has to be set within a context of alternative political scenarios.

Of course the future of Thatcherism does not depend upon having Mrs Thatcher as Prime Minister. One possible scenario would be the continuation or deepening of Thatcherism, perhaps under a different label. There is plenty of scope within the existing legislation for the further application of Thatcherite principles. The Secretary of State has the power to extend the coverage of UDCs and SPZs whether local authorities like it or not. Outside such areas the diminished powers of local plans and development control, together with the appeal system, provide plenty of scope for developers to impose their wishes. The extension of the Unitary Development Plan system and strategic guidance to the whole country would be easy to achieve and have a major centralising effect. There are also strong lobbies such as the Adam Smith Institute, British Property Federation, Royal Institute of Chartered Surveyors and the Institute of Economic Affairs, who are all putting forward ideas to push the government into more radical initiatives. These initiatives would further reduce the scope of planning and local democracy, for example, it has been suggested that development control should be removed from political influence, that Business Committees should be set up in local areas to consider planning matters, that the zoning system should be developed, and that greater delegation should be given to chief planning officers who would be appointed by central government.

It is often said (for example, King, 1987; Gamble, 1988) that Thatcherism has paved the way for a new consensus which would involve jettisoning some of the more extreme aspects of Thatcherism. However, this new world would be very different from that of post-war social democracy. The market would remain dominant and it would be impossible to reverse many of the actions of the Thatcher governments such as the privatisation programme or the sale of council houses. Are there aspects of planning that have been irreversibly changed as a result of Thatcherism? It is difficult to imagine that the politicisation of planning could be reversed. A return to the consensus of the post-war period with its vague and confused presentation of planning purpose is extremely unlikely. Thatcherism has introduced a greater sense of reality and destroyed the remaining myths planners might have held about their abilities to influence society. Britain has a capitalist system based on the market and property rights. However, this does not mean that there will be unquestioning acceptance of the dominance of these essential features and there will continue to be a reaction against their effects and implications. These different views will be more accentuated as they will not be cloaked under a general consensus about the value and purpose of planning. Planners will become more varied and more identified with either market, environmental or social goals. Each position will incorporate different attitudes, self-definition and areas of specialism. It also seems likely that the planning system will remain in the disaggregated form that has developed under Thatcherism, although maybe in a different configuration and for different reasons. This disaggregation would further accentuate the variation in planning practice.

Halting the forward march of Thatcherism

Thatcherism is reliant on economic success. To gain electoral support enough people have to feel that they benefit from the Thatcherite approach. This economic appeal is strengthened by that of strong leadership and decisive government. However, the ideology, built on economic liberalism and authoritarianism, is vulnerable if the electorate, or a sufficient proportion of it, lose faith in the ability of the government to come up with the economic or leadership goods. However, the Conservative Party has shown considerable flexibility in the past and there is no reason to suppose that they cannot devise a programme of Thatcherism with a human face, although this might expose its ideological contradictions. There is no doubt that they would continue to attack socialism and attempt to show that 'there is no alternative'. Thus it is important that any alternative inspires people with a different vision of the future. Such a vision needs to be based upon a clear and strong ideology while also being rooted in the reality of the present – it would need to explore the weaknesses of Thatcherism (Skidelsky, 1988). In terms of their importance for planning these weaknesses can be identified as:

1 the inefficiencies of the market;
2 the issue of the environment;
3 the attack on democracy.

In Chapter Two the antagonism between the processes of the market and planning intervention in the 'public interest' were discussed. Thatcherism can be viewed as a means of overcoming the contradictions that result from this antagonism by re-orienting planning to a market supportive role. There is no longer a co-existence of market and social democratic political processes. Instead the market is relied on to satisfy all needs and it is claimed that the spin-off effects will benefit everyone. The acceptance of inequalities makes the job of relying on the market easier. However, can the market match up to the task?

Although some individuals in Thatcher's Britain may be benefiting financially, there is no doubt that the public domain is suffering badly. For example, the 'overheating' of the South East caused by the free reign of market forces is causing severe problems in transport, motorways are congested and public transport overcrowded and under-resourced. Pollution, whether of the water supply or beaches or litter on the streets, is ever increasing. Open spaces in our towns and cities are constantly at risk. Market processes in housing have generated regional differentials affecting job mobility.

Such problems create economic inefficiencies for both individual enterprises and the economy as a whole. Greater competition from Europe after 1992 will expose these deficiencies even further. It is being suggested that this European challenge will create the need for a national infrastructure plan. Then there is the reliance on the spin-off effect to cater for social needs. Is there any evidence that developments, such as those in London's Docklands, do, in reality, generate a filter-down effect to everyone's benefit? Will the problems of the regions and inner areas be improved by this means? Can social stability be maintained with this strategy?

An alternative ideology could challenge these claims and attack the reliance on market criteria. Even Heseltine has been reported as stating that the market lacks morality. An ideological position could be developed that counters the self-interest and individualism of the market. It could stress caring, neighbourly and community values, the need to protect the public domain, and that the needs of some groups in society cannot be met through the market. Some social needs *may* be provided through the market process, for example, crèche facilities in supermarkets or for bank employees, but only when there is an economic reason such as greater sales or maintenance of a labour force. The decision is controlled by the imperatives of the market and the provision is likely to be constrained and insecure. Then, of course, there are many aspects that will not be covered in an individual entrepreneurial approach as the discussion of Houston indicated. For example, in this country we have seen in a dramatic way how difficult the market system finds the issue of safety. Whether one is a user of the underground or a woman who wishes to go out at night, safety in the public domain is an absolutely essential aspect of a civilised society. In this context the role that planning could play in satisfying social needs should be vigorously pursued and propagated.

The second weakness of Thatcherism is its difficulty in dealing with demands for the protection of the environment. Now of course the conflicts between development and the environment are not new – the balance between the two has always been a central issue for planning. What is significant about Thatcherism is the attempt to shift this balance towards a greater emphasis on development and economic imperatives. For most of the period since 1979 the increased conflicts created by this shift have been contained. However, the rise of the 'Green' issue and its popular support, as expressed in the European elections of 1989, creates a threat to this compromise. Thatcherism finds it difficult to incorporate these environmental demands. They give credence to an oppositional faction within the Conservative Party. At a more important and general level they open up the whole ideological question of the need for state intervention. It shows that there *is* more to society than individuals and their families and therefore challenges Thatcherism in a fundamental way. No doubt attempts will be made to absorb the issue into Thatcherism in a way that does not challenge its ideology – for example, concentrating on global or European scale issues or stressing the effects of environmental policies on the national economy.

The arrival of Chris Patten as Secretary of State for the Environment towards the end of 1989 signals an attempt by the government to counter the threat of the Green Movement. In his statements Patten has indicated that greater consideration should be given to environmental factors. This is illustrated in his attitude to the proposed new settlement at Foxley Wood and the revision of the Planning Policy Guidance Note on planning and housing. However, as already indicated there is a limit to how far these Green policies can be taken before they undermine the whole anti-interventionist ideology. Some of the Green demands could be met without creating such a threat. The 'not in my back yard' reaction to new development could be philosophically accommodated as an extension of property rights and incorporated under the umbrella of ameliorating the externalities of the market. The emphasis that Ridley and Patten have given to improving the quality of design in the new developments shows how they want developers to take greater care in reducing the detrimental effects of their schemes. However, there is considerable potential for tapping popular support for a wider definition of the

Green issue. An alternative 'populism' to that of Thatcherism could be invoked which links environmental protection to the fear of contamination of the fundamental elements of life: food and water. It would be very difficult to contain this wider view of the problem within an anti-interventionist ideology. Thus 'Green' issues could provide the basis on which to build a challenge to Thatcherism and would be of particular relevance in demonstrating the need for planning.

The third weakness of Thatcherism is its disdain for democracy and participation. Patten has shown signs of softening this attitude in stating the importance of the local community and local authority in deciding on new housing developments. However, again there are limits on how far this can be taken without detracting from centralised government control which is a fundamental aspect of Thatcherism. An alternative ideology could demonstrate the benefits from a more participatory style of government. Popular pressure for greater involvement existed before 1979 and this could be rekindled. In recent years there has been increasing discontent over the authoritarian nature of Thatcher's governments. Post-war social democracy incorporated both political rights in terms of democratic and participatory principles, and social rights in terms of certain basic levels of welfare provision. Thatcherism has attacked both these rights. Political and democratic processes have been replaced by a reliance on market choice, consumer sovereignty, and centralisation while social rights are no longer seen as the responsibility of government. There has been a growing campaign, centred around Charter 88, for a resurrection and formalisation of citizen rights (see also Krieger, 1986, pp. 209–13).

Such a movement towards the re-establishment of democratic principles and basic rights could affect issues relating to development and the environment. Can practices of the past, such as popular planning or decentralisation, be reintroduced or do new approaches have to be found? How is the tension resolved between maximum involvement and the need to maintain certain universal principles, such as avoiding discrimination, which implies some form of centralised monitoring (Brownill, 1988)? Planning should build and expand on its track record of participation.

Amongst the reactions to Thatcherite urban planning over the last decade there have been two rather opposing factions. The pragmatic approach suggests that much can be achieved through a change of attitude to planning. Once a caring and redistributive philosophy has been established, Thatcherism can be turned around without the need for major legislation (Thompson, 1990). This 'new realist' approach stresses that the market has to be accepted as the essential mechanism of society and that new ways must be found to mould it to human needs. According to this view a future programme would be based on adapting or changing the existing planning machinery through, for example, repealing unwanted Circulars, ensuring that all bodies are democratic, increasing public involvement, giving greater importance to social and environmental criteria and extending joint venture, partnership and planning gain approaches. The second faction condemns the 'new realism' and believes that such negotiation or bargaining will result in planning decisions always being made on developers' terms (Colenutt, 1990). It implies the removal from the agenda of basic issues such as who gains and who loses, who owns land and who controls the development process. Partnership means forcing local authorities into creating the right climate for development so they have to adopt a purely defensive stance. This defensiveness makes alliances with community groups very difficult. According to this second approach a future programme must be built around establishing community control and involvement. While there are such divisions in the formulation of an alternative the opportunity exists for Thatcherism to adapt and march onward. There is a need to find common ground between such factions and link this to a broad movement of popular support.

Planners, whether in education or practice, have an important role to play in ensuring that there is an alternative. Although a new ideological and political climate is a necessary precondition for change, it is important for planners, who believe in a broader purpose for planning than that prescribed under Thatcherism, to ensure that the wider aims and the means of achieving them are kept on the agenda (Montgomery and Thornley, 1990). In education this means keeping the debates open and not getting sucked into a purely pragmatic response to contemporary conditions. In practice it means exploring every loophole and developing and publicising any initiatives that fulfil wider aims. The virtues of a strong and imaginative planning system need to be clearly expressed and propagated.

Bibliography

Abercrombie, N., Hill, S. and Turner, B. (1980), *The Dominant Ideology Thesis*, London, Allen & Unwin.

Adam Smith Institute (ASI) (1983), *Omega Report; Local Government Policy*, London, Adam Smith Institute.

Adcock, B. (1984), 'Regenerating Merseyside Docklands', *Town Planning Review*, Vol. 55, No. 3.

Allison, L. (1986), 'What is urban planning for?', *Town Planning Review*, Vol. 57, No. 1.

Ambrose, P. (1986), *Whatever Happened to Planning*, London, Methuen.

Ambrose, P. and Colenutt, B. (1975), *The Property Machine*, Harmondsworth, Penguin.

Anderson, J. (1980), 'Back to the 19th century', *New Statesman*, July 11th.

Anderson, J. (1983), 'Geography as ideology and the politics of crisis: the EZ experiment', in Anderson, J., Duncan, S. and Hudson, R., *Redundant Spaces in Cities and Regions*, London, Academic Press.

Anderson, J., Smith, D. and Walker R. (1981), 'Heseltine's Excalibur', *Planning*, July 10th.

Angell, R. and Taylor N. (1985), 'Unitary Development Plans: an initial appraisal', *The Planner*, Dec.

Arnold-Baker, C. (1981), *Local Government, Planning and Land Act 1980*, London, Butterworth.

Ashworth, W. (1954), *The Genesis of Modern British Town Planning*, London, Routledge & Kegan Paul.

Aughey, A. (1983), 'Mrs Thatcher's philosophy', *Parliamentary Affairs*, Vol. 36, No. 4.

Aughey, A. (1984), *Elements of Thatcherism*, University of Southampton, PSA Conference Paper, Apr. 3–5.

Backwell, J. and Dickens, P. (1978), *Town Planning, Mass Loyalty and the Restructuring of Capital: The Origins of the 1947 Planning Legislation Revisited*, University of Sussex, Urban and Regional Studies Working Paper No. 11.

Bailey, R. (1982), 'Public enterprise', *New Society*, Jan. 21st.

Ball, M. (1983), *Housing Policy and Economic Power*, London, Methuen.

Banfield, E. (1971a), 'Putting social science to work is a risky undertaking',

Social Science Quarterly, Vol. 51, No. 4.

Banfield, E. (1971b), 'In reply', *Trans-action*, Vol. 8, No. 5/6.

Banfield, E. (1974), *Unheavenly City Revisited*, Boston, Mass., Little Brown and Co.

Banham, R., Barker, P., Hall, P. and Price, C. (1969), 'Non-plan: an experiment in freedom', *New Society*, Mar. 20th.

Barnes, I. and Preston, J. (1985), 'The Scunthorpe EZ: an example of muddled interventionism', *Public Administration*, Vol. 63, Summer.

Barnes, J. (1988), *Lessons from the LDDC*, Polytechnic of Central London, Radical Planning Initiative Conference Paper, Jan., reproduced in Montgomery, J. and Thornley, A. (eds) (1990), *Radical Planning Initiatives; New directions for Urban Planning in the 1990s*, Aldershot, Gower.

Barry, N., Burton, J., Gissurarson, H.H., Gray, J., Shearmur, I. and Vaughn, K.I. (1984), *Hayek's Serfdom Revisited*, London, Institute of Economic Affairs.

Beattie, A. (1979), 'Macmillan's mantle: the Conservative Party in the 1970s', *Political Quarterly*, Vol. 50, No. 3.

Behrens, R. (1980), *The Conservative Party from Heath to Thatcher*, Farnborough, Saxon House.

Belsey, A. (1986), 'The New Right, social order and civil liberties', in Levitas, R. (ed.), *The Ideology of the New Right*, Cambridge, Polity Press.

Benton, S. (1987), 'Death of a citizen', *New Statesman*, Vol. 114, No. 2,956, Nov 20th.

Billingham, D. (1981), *The Creation of the LDDC*, Polytechnic of Central London Planning Studies No. 13.

Blake, R. (1985), *The Conservative Party from Peel to Thatcher*, London, Fontana.

Bleaney, M. (1983), 'Conservative economic strategy', in Hall, S. and Jacques, M. (eds), *The Politics of Thatcherism*, London, Lawrence & Wishart.

Blowers, A. (1980), *The Limits of Power: The Politics of Local Planning Policy*, Oxford, Pergamon.

Blowers, A. (1986), 'Town planning – paradoxes and prospects', *The Planner*, April.

Boaden, N. (1982), 'Urban Development Corporations – threat or challenge?', *Local Government Studies*, Vol. 8, No. 4.

Boddy, M. and Fudge, C. (eds) (1984), *Local Socialism?*, London, Macmillan.

Bogdanor, V. (1983), 'A deep transformation? The meaning of Mrs Thatcher's victory', *Encounter*, Sept./Oct.

Bond, R. (1980), 'A man of enterprise', *Surveyor*, July 3rd.

Bosanquet, N. (1983), *After the New Right*, London, Heinemann.

Botham, R. and Lloyd, G. (1983), 'The political economy of Enterprise Zones', *National Westminster Bank Quarterly Review*, May.

Boyle, R. (1983), 'British export held back by Uncle Sam', *Planning*, Feb. 18th.

Boyson, R. (ed.) (1970), *Right Turn*, London, Churchill Press.

Boyson, R. (ed.) (1971), *Down with the Poor*, London, Churchill Press.

Bracewell-Milnes, B. (1974), 'Market control over land use "planning"', in Walters, A. *et al.*, *Government and the Land*, London, Institute of Economic Affairs.

Bracken, I. (1982), 'Problems and issues in structure plan review and alteration', *The Planner*, Jan./Feb.

Brand, C.M. and Williams, D.W. (1984), 'Circular 22/80: three years on', *Estates Gazette*, Feb. 18th.

Brindley, T., Rydin, Y. and Stoker, G. (1989), *Remaking Planning*, London, Unwin Hyman.

Bristow, M.R. (1985a), 'How unitary is Unitary? Some comments on the new British Unitary Plan system', *Built Environment*, Vol. 11, No. 3.

Bristow, M.R. (1985b), 'Some questions on Unitary Development Plans – a plain man's guide?', *Regional Studies*, Vol. 19, No. 3.

British Property Federation (BPF) (1986), *The Planning System – A Fresh Approach*, London, British Property Federation.

Brittan, S. (1980), 'Hayek, the New Right and the crisis of social democracy', *Encounter*, pp. 31–46.

Broadbent, A. (1977), *Planning and Profit in the Urban Economy*, London, Methuen.

Bromley, R. and Morgan, R. (1985), 'The effects of enterprise zone policy – evidence from Swansea', *Regional Studies*, Vol. 19, No. 5.

Brookes, J. (1989), 'Cardiff Bay Renewal Strategy – another hole in the democratic system', *The Planner*, Jan.

Brownill, S. (1988), 'The People's Plan for the Royal Docks: some contradictions in popular planning', *Planning Practice and Research*, No. 4 Spring, reproduced in Montgomery, J. and Thornley, A. (eds) (1990), *Radical Planning Initiatives*, Aldershot, Gower.

Brownill, S. (1989), 'Dealing in Docklands', *Planning*, Apr. 21st.

Bruce, A. (1980), 'Structure plan amendments show up centralist tendencies', *Planning*, Nov. 14th.

Bruce, M. (1968), *The Coming of the Welfare State*, London, Batsford.

Bruton, M. (1983), *Legislation and the Role of the Town Planner in Society*, University of Wales Institute of Science and Technology, Department of Town Planning Papers in Planning Research No. 61.

Bruton, M. and Nicholson, D. (1985), 'Local planning in practice: a review', *The Planner*, Dec.

Bruton, M. and Nicholson, D. (1987), *Local Planning in Practice*, London, Hutchinson.

Butler, E. (1983), *Hayek*, London, Temple Smith.

Butler, S.M. (1981), *EZs: Greenlining the Inner Cities*, London, Heinemann.

Byrne, S.P. (1984), 'Planning gain – the local authority's view', in *Contemporary Planning Policies*, Journal of Planning and Environment Law Occasional Papers, London, Sweet & Maxwell.

Cabinet Office (1988), *Action for Cities*, London, HMSO.

Cadman, D. (1981), 'Urban change, EZs and the role of investors', *Built Environment*, Vol. 7, No. 1.

Cadman, D. (1983), 'Planning – who needs it?', *The Planner*, November/December.

Cawson, A. (1977), *Environmental Planning and the Politics of Corporatism*, University of Sussex, Urban and Regional Studies Working Paper No.7.

Cawson, A. (1986), *Corporatism and Political Theory*, Oxford, Blackwell.

Centre for Policy Studies (1980), *A Bibliography of Freedom*, London, Centre for Policy Studies (second edition).

Centre for Policy Studies (1985), *Comments on 'The City of London Draft Local Plan' of November 1984*, London, Centre for Policy Studies.

Cherry, G. (1972), *Urban Change and Planning: A History of Urban Development in Britain since 1750*, Henley-on-Thames, Foulis.

Cherry, G. (1974), *The Evolution of British Town Planning*, Leighton Buzzard, Leonard Hill.

Cherry, G. (1982), *The Politics of Town Planning*, London, Longman.

Cheshire, P. and Evans, A. (1981), 'Opportunity for developers', *Estates Gazette*, Feb. 21st.

City of London (1984), *The City of London Draft Local Plan*, London, City of London.

Clarke, G. and Dear, M. (1984), *State Apparatus: Structures and Language of Legitimacy*, Winchester, Mass., Allen & Unwin.

Coates, D. and Hillard, J. (1986), *The Economic Decline of Modern Britain*, Brighton, Wheatsheaf.

Cohen, G., Bosanquet, N., Ryan, A., Parekh, B., Keegan, W. and Gress, F. (1986), *The New Right. Image and Reality*, London, The Runnymede Trust.

Colenutt, B. (1981), 'The national interest', *Town and Country Planning*, Vol. 50, No. 4.

Colenutt, B. (1987), 'The challenge of UDCs', *Capital Issues*, Aug.

Colenutt, B. (1990), 'Politics, not partnership', in Montgomery, J. and Thornley, A. (eds), *Radical Planning Initiatives*, Aldershot, Gower.

Cooke, P. (1983), *Theories of Planning and Spatial Development*, London, Hutchinson.

Corfield, F. (1984), 'The changing role of central government in planning policy and objectives', in *Contemporary Planning Policies*, Journal of Planning and Environment Law Occasional Papers, London, Sweet & Maxwell.

Counter Information Services (1973), *The Recurrent Crisis of London: Anti-Report on the Property Developers*, London, CIS.

Cowling, M. (ed.) (1978), *Conservative Essays*, London, Cassell.

Cox, A. (1980), 'Continuity and discontinuity in conservative urban policy', *Urban Law and Policy*, Vol. 3.

Cox, A. (1981), 'Adversary politics and land policy', *Political Studies*, Vol. 29.

Cox, A. (1984), *Adversary Politics and Land*, Cambridge, Cambridge University Press.

Crosland, A. (1956), *The Future of Socialism*, London, Cape.

Cullingworth, J.B. (1975), *Environmental Planning, 1939–69. Volume I: Reconstruction and Land Use Planning*, London, HMSO.

Cullingworth, J.B. (1980), *Environmental Planning, 1939–69. Volume IV: Land Values, Compensation and Betterment*, London, HMSO.

Darke, R. (1979), 'Public participation and state power: the case of South Yorkshire', *Policy and Politics*, Vol. 7, No. 4.

Davies, H.W.E. (1983), 'Development control in the 1980s', *The Planner*, Feb.

Davies, H.W.E., Edwards, D. and Rowley, A.R. (1986), 'The relationship between Development Plans, development control and appeals', *The Planner*, Oct.

Davis-Coleman, C. (1987), 'The pros and cons of the UDCs', *Municipal Journal*, Feb. 6th.

Deakin, N. (1987), *The Politics of Welfare*, London, Methuen.

Debenham, Tewson and Chinnocks (1986), *A Preliminary Review of the Draft Local Plan for the City of London*, London, Debenham, Tewson & Chinnocks.

Debenham, Tewson and Chinnocks (1987), *Special Report on the Town and Country Planning (Use Classes) Order 1987*, London, Debenham, Tewson & Chinnocks.

Delafons, J. (1969), *Land Use Controls in the United States*, Cambridge, Mass., MIT Press.

Denman, D.R. (1980), *Land in a Free Society*, London, Centre for Policy Studies.

Denman, S.E. (1974), 'Present policies – aims and results', in Walters, A. *et al.*, *Government and the Land*, London, Institute of Economic Affairs.

Denyer-Green, B. (1988), 'Housing and Planning Act; SPZ and local plan provisions', *Estates Gazette*, Mar. 12th.

Department of Employment (1986), Cmnd 9794, *Building Business – Not Barriers*, London, HMSO.

Department of Environment (DOE) (1979), Consultation paper, *Urban Development Corporations*, London, Department of Environment.

Department of Environment (1980a), Circular 9/80, *Land for Private Housebuilding*, London HMSO.

Department of Environment (1980b), Circular 22/80, *Development Control – Policy and Practice*, London, HMSO.

Department of Environment (1981a), *Urban Development Corporations – Summary of Legislative Provisions*, London, Department of Environment.

Department of Environment (1981b), Memorandum, *The Proposed LDDC: Planning and Development Control*, London, Department of Environment.

Department of Environment (1981c), Circular 23/81, *Local Government, Planning and Land Act 1980. Town and Country Planning: Development Plans*, London, HMSO.

Department of Environment (1981d), *London Docklands Development Corporation [Planning Functions] Order*, London, HMSO.

Department of Environment (1983a), Circular 22/83, *Town and Country Planning Act 1971: Planning Gain*, London, HMSO.

Department of Environment (1983b), Internal paper, *Local Development Schemes*, March.

Department of Environment (1983c), Internal paper, *EZ Schemes*, June.

Department of Environment (1983d), *Advice on the EZ Planning System in England and Wales*, London, Department of Environment.

Department of Environment (1984a), Circular 16/84, *Industrial Development*, London, HMSO.

Department of Environment (1984b), Circular 22/84, *Memorandum on Structure and Local Plans*, London, HMSO.

Department of Environment (1984c), Consultation paper, *Simplified Planning Zones*, London, Department of Environment.

Bibliography

Department of Environment (1985a), Cmnd 9571, *Lifting the Burden*, White Paper, London, HMSO.
Department of Environment (1985b), Circular 1/85, *The Use of Conditions in Planning Permissions*, London, HMSO.
Department of Environment (1985c), Circular 14/85, *Development and Employment*, London, HMSO.
Department of Environment (1985d), Circular 30/85, *Local Government Act 1985: Sections 3, 4 and 5: Schedule 1 Town and Country Planning: Transitional Matters*, London, HMSO.
Department of Environment (1985e), Circular 31/85, *Aesthetic Control*, London, HMSO.
Department of Environment (1985f), *Simplified Planning Zones; Revised Proposals*, London, Department of Environment.
Department of Environment (1986a), *Proposals to Modernise the Town and Country Planning (Use Classes) Order 1972*, London, Department of Environment.
Department of Environment (1986b), Consultation paper, *The Future of Development Plans*, London, Department of Environment.
Department of Environment (1986c), Circular 2/86, *Development by Small Businesses*, LOndon, HMSO.
Department of Environment (1987a), *Town and Country Planning (Use Classes) Order 1987*, London, HMSO.
Department of Environment (1987b), Circular 2/87, *Award of costs incurred in planning and compulsory purchase order proceedings*, London, HMSO.
Department of Environment (1987c), Circular 13/87, Change of Use of Buildings and Other Land: the Town and Country Planning (Use Classes) Order, London, HMSO.
Department of Environment (1987d), Circular 25/87, *Housing and Planning Act 1986: Town and Country Planning: Simplified Planning Zones*, London, HMSO.
Department of Environment (1988a), *Planning Policy Guidance No. 1: General Policy and Principles*, London, HMSO.
Department of Environment (1988b), *Planning Policy Guidance No. 4: Small Businesses*, London, HMSO.
Department of Environment (1988c), *Planning Policy Guidance No. 5: Simplified Planning Zones*, London, HMSO.
Department of Environment (1988d), *Planning Policy Guidance No. 12: Local Plans*, London, HMSO.
Department of Environment (1988e), Circular 22/88, *General Development Order 1988 Consolidation*, London, HMSO.
Department of Environment (1989), Cm. 569, *The Future of Development Plans*, White Paper, London, HMSO.
Department of Trade and Industry (1985), *Burdens on Business: Report of a Scrutiny of Administrative and Legislative Requirements*, London HMSO.
Dobson, M. (1981), 'Housing on condition', *Planning*, Sept. 25th.
Docklands Consultative Committee (1988), *Urban Development Corporations – Six Years in London's Docklands*, London, DCC.

Donnison, D. and Soto, P. (1980), *The Good City: A Study of Urban Development and Policy in Britain*, London, Heinemann.

Drucker, H., Dunleavy, P., Gamble, A. and Peele, G. (eds), (1983), *Developments in British Politics*, London, Macmillan.

Duffy, H. (1987), 'The two-faced phoenix', *Financial Times*, May 13th.

Duncan, S.S. and Goodwin, M. (1982), 'The local state and restructuring social relations. Theory and practice', *International Journal of Urban and Regional Research*, Vol. 6.

Duncan, S.S. and Goodwin, M. (1988), *The Local State and Uneven Development*, Cambridge, Polity Press.

Dunleavy, P. (1981), 'Professions and policy change: notes towards a model of ideological corporatism', *Public Administration Bulletin*, Vol. 36, pp. 3–16.

Dunleavy, P. and O'Leary, B. (1987), *Theories of the State*, London, Macmillan.

Dyson, A. (1970), 'Farewell to the left', in Boyson, R. (ed.), *Right Turn*, London, Churchill Press.

Eccleshall, R. (1977), 'English conservatism as ideology', *Political Studies*, Vol. 25, No. 1.

Eccleshall, R. (1984), 'Introduction: the world of ideology', 'Liberalism', and 'Conservatism', in Eccleshall, R., Geoghegan, V., Jay, R. and Wilford, R., *Political Ideologies*, London, Hutchinson.

Eccleshall, R., Geoghegan, V., Jay, R. and Wilford, R. (1984), *Political Ideologies*, London, Hutchinson.

Edgar, D. (1983), 'Bitter harvest', *New Socialist*, Sept./Oct.

Edgar, D. (1986), 'The free or the good', in Levitas, R. (ed.), *The Ideology of the New Right*, Cambridge, Polity Press.

Edwards, A.E., Leslie, J., O'Donovan, G. and Carter, L.C. (1986), 'Simplified Planning Zones – a reaction to the proposal', *Planning Outlook*, Vol. 29, No.1.

Ehrman, R. (1988), *Planning Planning*, London, Centre for Policy Studies.

Elson, M. (1986), *Green Belts: Conflict Mediation in the Urban Fringe*, London, Heinemann.

Elton, Lord (1986a), 'Address to the RTPI Summer School 1985', *The Planner*, Feb.

Elton, Lord (1986b), 'The agenda for strategic guidance in London', letter to London Planning Advisory Committee, Department of Environment, July 30th.

Estates Gazette (1980), 'Editorial – Glimpse of freedom', *Estates Gazette*, Dec. 6th.

Estates Gazette (1981), 'Editorial – Zones and reservations', *Estates Gazette*, June 20th.

Eversley, D. (1973), *The Planner in Society*, London, Faber.

Fallows, A. (1983), 'Compensation', in *Structure Plans and Local Plans – Planning in Crisis*, Journal of Planning and Environment Law Occasional Papers, London, Sweet & Maxwell.

Faludi, A. (ed.) (1973), *A Reader in Planning Theory*, Oxford, Pergamon.

Feagin, J.R. (1988), *Free Enterprise City: Houston in Political-Economic Perspective*, New Brunswick, NJ, Rutgers University Press.

Flynn, N. (1989), 'The 'New Right' and social policy', *Policy and Politics*, Vol. 17, No. 2.

Foley, D. (1960), 'British town planning: one ideology or three?', *British Journal of Sociology*, Vol. 11, pp. 211–31. Also in Faludi, A. (ed.) (1973), *A Reader in Planning Theory*, Oxford, Pergamon.

Fraser, D. (1973), *The Evolution of the British Welfare State*, London, Macmillan.

Friedman, J. (1971), 'Unheavenly city', *American Institute of Planners Journal*, March.

Friedman, M. (1962), *Capitalism and Freedom*, Chicago, University of Chicago Press.

Friedman, M. and Friedman, R. (1980), *Free to Choose*, Harmondsworth, Penguin.

Gamble, A. (1974), *The Conservative Nation*, London, Routledge and Kegan Paul.

Gamble, A. (1979a), 'The free economy and the strong state', in Miliband, R. and Saville, J. (eds), *The Socialist Register*, London, Merlin Press.

Gamble, A. (1979b), 'The Conservative Party', in Drucker, H.M. (ed.), *Multi-Party Britain*, London, Macmillan.

Gamble, A. (1981), *Britain in Decline*, London, Macmillan.

Gamble, A. (1983), 'Thatcher: the second coming', *Marxism Today*, July.

Gamble, A. (1984), 'This lady's not for turning: Thatcherism Mark III', *Marxism Today*, July.

Gamble, A. (1988), *The Free Economy and the Strong State*, London, Macmillan.

Garner, J.F. (1985), 'The decline of planning control', *Journal of Planning and Environment Law*, Nov.

Geertz, C. (1964), 'Ideology as a cultural system', in Apter, D. (ed.), *Ideology and Discontent*, New York, Free Press of Glencoe.

George, V. and Wilding, P. (1976), *Ideology and Social Welfare*, London, Routledge & Kegan Paul (updated edition, 1985).

Gilmour, I. (1977), *Inside Right: A Study of Conservatism*, London, Hutchinson.

Gilmour, I. (1980), 'Foreword', in Layton-Henry, Z. (ed.), *Conservative Party Politics*, London, Macmillan.

Glass, R. (1959), 'The evaluation of planning: some sociological considerations', *International Social Science Journal*, Vol. 11, pp. 393–409. Also in Faludi, A. (ed.) (1973), *A Reader in Planning Theory*, Oxford, Pergamon.

Goldsmith, W. (1982), 'EZs: if they work, we're in trouble', *International Journal of Urban and Regional Research*, Vol. 6, No. 3.

Goodchild, R.N. and Denman, D.R. (1988), *Planning Fails the Inner Cities*, London, Social Affairs Unit.

Gough, I. (1979), *The Political Economy of the Welfare State*, London, Macmillan.

Graham, D. and Clarke, P. (1986), *The New Enlightenment: The Rebirth of Liberalism*, London, Macmillan.

Grant, M. (1982), *Urban Planning Law*, London, Sweet & Maxwell.

Grant, M. and Healey, P. (1985), 'The rise and fall of planning', in Loughlin,

M., Gelfand, M.D. and Young, K. (eds), *Half a Century of Municipal Decline, 1935–85*, London, Allen & Unwin.

Greater London Council (GLC) (1984), *London Docklands: Review of the First Two Years Operation of the LDDC*, London, Greater London Council.

Greater London Council (1985), *Erosion of the Planning System*, London, Greater London Council.

Green, D.G. (1987), *The New Right*, Brighton, Wheatsheaf.

Gregg, P. (1967), *The Welfare State*, London, G. Harrap & Co.

Griffiths, R. (1986), 'Planning in retreat? Town planning and the market in the 1980s', *Planning Practice and Research*, No. 1.

Gyford, J. (1985), *The Politics of Local Socialism*, London, Allen & Unwin.

Hadley, G. (1984), 'EZs in Britain: the form and consequences of the planning scheme approach', *Planning Outlook*, Vol. 27, No. 1.

Hague, C. (1984), *The Development of Planning Thought*, London, Hutchinson.

Hajer, M.A. (1989), *City Politics*, Aldershot, Gower.

Hall, P. (1977), 'Green Fields and Grey Areas', RTPI Annual Conference Paper June 15th, London, Royal Town Planning Institute.

Hall, P. (1982), 'EZs: a justification' and 'Response', *International Journal of Urban and Regional Research*, Vol. 6, No. 3.

Hall, P. (1983a), 'Enterprise Zones and freeports revisited', *New Society*, Mar. 24th.

Hall, P. (1983b), 'Housing, planning, land and local finance: the British experience', *Urban Law and Policy*, Vol. 6.

Hall, P. (1984), 'Enterprises of great pith and moment?', *Town and Country Planning*, Vol. 53, No. 11.

Hall, P., Gracey, H., Drewett, R. and Thomas, R. (1973), *The Containment of Urban England*, London, Allen & Unwin.

Hall, S. (1983), 'The great moving right show', in Hall, S. and Jacques, M. (eds), *The Politics of Thatcherism*, London, Lawrence & Wishart.

Hall, S. (1985), 'Authoritarian populism: a reply', *New Left Review*, No. 151.

Hall, S. (1988), *The Hard Road to Renewal*, London, Verso.

Hall, S. and Jacques, M. (eds) (1983), *The Politics of Thatcherism*, London, Lawrence & Wishart.

Hambleton, R. (1986), *Rethinking Policy Planning*, University of Bristol, School of Advanced Urban Studies.

Hardy, D. and Ward C. (1984), 'Lessons from the plotlands', *The Planner*, Nov.

Harris, N. (1971), *Beliefs in Society*, Harmondsworth, Penguin.

Harris, N. (1972), *Competition and the Corporate Society*, London, Methuen.

Harrison, B. (1982), 'The politics and economics of the urban enterprise zone proposal', *International Journal of Urban and Regional Research*, Vol. 6, No. 3.

Harrison, M.L. (ed.), (1984), *Corporatism and the Welfare State*, Aldershot, Gower.

Harrison, M.L. (1987), 'Property rights, philosophies, and the justification of planning control', in Harrison, M.L. and Mordey, R., *Planning control: Philosophies, Prospects and Practice*, Beckenham, Croom Helm.

Bibliography

Harrison, M.L. and Mordey, R. (1987), *Planning Control: Philosophies, Prospects and Practice*, Beckenham, Croom Helm.
Hart, L. (1983), 'Liberal Political Economy and Planning 1979–83', Unpublished M.Phil dissertation, University of Newcastle-upon-Tyne.
Harvey, D. (1985a), *Consciousness and the Urban Experience*, Oxford, Blackwell.
Harvey, D. (1985b), *The Urbanisation of Capital*, Oxford, Blackwell.
Hayek, F.A. (1944), *The Road to Serfdom*, London, Routledge & Kegan Paul.
Hayek, F.A. (1960), *The Constitution of Liberty*, London, Routledge & Kegan Paul, reprinted 1990 by Routledge.
Hayek, F.A. (1982), *Law, Legislation, and Liberty*, London, Routledge & Kegan Paul, (first published in three volumes in 1973, 1976, 1979).
Hayek, F.A. (1984), 'The moral tradition that reason must recognise', *Guardian*, Sept. 17th.
Heald, D. (1983), *Public Expenditure*, Oxford, Martin Robertson.
Healey, P. (1983), *Local Plans in British Land Use Planning*, Oxford, Pergamon.
Healey, P. (1986), 'The role of Development Plans in the British planning system: an empirical assessment', *Urban Law and Policy*, Vol. 8.
Healey, P. (1987), 'The future of local planning and development control', *Planning Outlook*, Vol. 30. No. 1.
Healey, P., McNamara, P., Elson, M. and Doak, A. (1989), *Land Use Planning and the Mediation of Urban Change*, Cambridge, Cambridge University Press.
Hebbert, M. (1977), 'The evolution of British town and country planning', University of Reading PhD, unpublished.
Held, D. (1984), 'Power and legitimacy in contemporary Britain', in McLennan, G., Held, D. and Hall, S. (eds), *State and Society in Contemporary Britain*, Cambridge, Polity Press.
Henneberry, J. (1985), 'The Use Classes Order and high tech developments', *The Planner*, Sept.
Heseltine, M. (1979), 'Secretary of State's Address', *Report of Proceedings RTPI Summer School 1979*, London, Royal Town Planning Institute.
Heseltine, M. (1982), 'Secretary of State's Address to the RTPI Summer School 1981', *The Planner*, Feb.
Hindess, B. (1987), *Freedom, Equality and the Market*, London, Tavistock.
Holt, G. (1985), 'Class of '86 shapes up', *Planning*, Dec. 13th.
Holt, G. (1986), 'Circular tips balance – but only a little', *Planning*, Feb. 21st.
Home, R.K. (1987), 'Planning decision statistics in the Use Classes debate', *Journal of Planning and Environment Law*, March.
Howe, G. (1978), *A Zone of Enterprise to Make all Systems 'Go'*, Speech to Bow Group June 26th and published by Conservative Central Office.
Howe, G. (1982), *Conservatism in the Eighties*, London, Conservative Political Centre.
Howell, D. (1980), *The Conservative Tradition and the 1980s*, London, Centre for Policy Studies.
Humber, R. (1986), 'New urban initiatives', *House Builder*, Vol. 45, No. 2, Feb.

Jacobs, J. (1965), *The Death and Life of Great American Cities*, Harmondsworth, Penguin, (first published in the USA by Random House in 1961).

Jacobs, J. (1970), *The Economy of Cities*, London, Jonathan Cape.

Jenkin, P. (1984), 'Secretary of State's Address to the RTPI Summer School 1983', *The Planner*, Feb.

Jessop, B. (1982), *The Capitalist State*, Oxford, Blackwell.

Jessop, B., Bonnett, K., Bromley, S. and Ling, T. (1984), 'Authoritarian populism, two nations, and Thatcherism', *New Left Review*, No. 147.

Jessop, B., Bonnett, K., Bromley, S. and Ling, T. (1988), *Thatcherism*, Cambridge, Polity Press.

Job, S. (1984), 'The Enterprise Zones: justification for SPZs', paper given to Town and Country Planning Association Conference, 'From EZ to SPZ: Lessons from the Enterprise Zones'.

Johnston, B. (1983a), 'Relaxing on a rates holiday', *Planning*, May 6th.

Johnston, B. (1983b), 'A step in no direction from "bland" gain advice', *Planning*, May 20th.

Johnston, B. (1985), 'David tackles planning Goliath?', *Planning*, July 26th.

Johnston, B. (1986), 'World of difference in world of property', *Planning*, Nov. 7th.

Johnston, B. (1987a), 'The Future of Development Plans: adequate thinking gets miss in paper', *Planning*, Jan. 30th.

Johnston, B. (1987b), 'The Future of Development Plans: district chiefs lay claim to single tier', *Planning*, Feb. 6th.

Jones, P. (1980), 'The Thatcher experiment: tensions and contradictions in the first year', in *Politics and Power No. 2*, London, Routledge & Kegan Paul.

Jones, R. (1982), *Town and Country Chaos; a Critical Analysis of Britain's Planning System*, London, Adam Smith Institute.

Joseph, K. (1976a), *Monetarism is Not Enough*, London, Centre for Policy Studies.

Joseph, K. (1976b), *Stranded in the Middle Ground*, London, Centre for Policy Studies.

Joseph, K. (1978), *Conditions for Fuller Employment*, London, Centre for Policy Studies.

Joseph, K. and Sumption, J. (1979), *Equality*, London, John Murray.

Journal of Planning and Environment Law (1981), 'Current topics', *Journal of Planning and Environment Law*, Nov.

Journal of Planning and Environment Law (1982a), 'Current topics', *Journal of Planning and Environment Law*, Jan.

Journal of Planning and Environment Law (1982b), 'Planning gain: the Law Society's observations', *Journal of Planning and Environment Law*, June.

Jowell, J. (1983), 'Structure plans and social engineering', in *Structure Plans and Local Plans – Planning in Crisis*, Journal of Planning and Environment Law Occasional Papers, London, Sweet & Maxwell.

Jowell, J. and Grant, M. (1983), 'Guidelines for planning gain', *Journal of Planning and Environment Law*, July.

Jowell, J. and Noble, D. (1980), 'Planning as social engineering: notes on the first English structure plans', *Urban Law and Policy*, Vol. 3.

Karski, A. (1986), 'Drawing the sting in the tail of White Paper', *Planning*, Feb. 21st.

237

Bibliography

Kavanagh, D. (1987), *Thatcherism and British Politics*, Oxford, Oxford University Press.

Keating, M. (1981), 'From rhetoric to reality', *Municipal Journal*, Aug. 21st.

Keating, M., Midwinter, A. and Taylor, D. (1984), 'Enterprise Zones: implementing the unworkable', *Political Quarterly*, Jan./Mar.

Keegan, W. (1984), *Mrs Thatcher's Economic Experiment*, London, Allen Lane.

King, D.S. (1987), *The New Right: Politics, Markets and Citizenship*, London, Macmillan.

Kirby, D. and Holt G. (1986), 'Planning responses to non-retail uses in shopping centres', *The Planner*, July.

Kirk, G. (1980), *Urban Planning in a Capitalist Society*, London, Croom Helm.

Klosterman, R.E. (1985), 'Arguments for and against planning', *Town Planning Review*, Vol. 56, No. 1.

Knibbs, T. (1989), 'An open approach to community benefit', *Planning*, Aug. 18th.

Krieger, J. (1986), *Reagan, Thatcher, and the Politics of Decline*, Cambridge, Polity Press.

Kristol, I. (1970), 'The cities: a tale of two classes', *Fortune*, July.

Kristol, I. (1978), *Two Cheers for Capitalism*, New York, Basic Books.

Lambeth London Borough (1984), *Town Planning Committee Paper TP110/84–85*, London, Lambeth LB.

Lambeth London Borough (1985), *Town Planning Committee Paper TP103/85–86*, London, Lambeth LB.

Lampman, R. (1971), 'Moral realism and the poverty question', *Social Science Quarterly*, Vol. 51, No. 4.

Lawless, P. (1983), Section on Planning in 'Parties, policy and the election. One: the Tories', *Critical Social Policy*, Vol. 8, pp. 34–5.

Lawless, P. (1986), *The Evolution of Spatial Policy*, London, Pion.

Lawson, N. (1980), *The New Conservatism*, London, Centre for Policy Studies.

Lawson, N. (1981), *Thatcherism in Practice*, London, Conservative Political Centre.

Layton-Henry, Z. (ed.) (1980), *Conservative Party Politics*, London, Macmillan.

Lejeune, A. (1970), 'Killing the geese', in Boyson, R. (ed.), *Right Turn*, London, Churchill Press.

Levitas, R. (1986a), 'Ideology and the New Right' and 'Competition and compliance: the utopias of the New Right', in Levitas, R. (ed.), *The Ideology of the New Right*, Cambridge, Polity Press.

Levitas, R. (ed.) (1986b), *The Ideology of the New Right*, Cambridge, Polity Press.

Leys, C. (1983), *Politics in Britain*, London, Heinemann.

Lloyd, M.G. (1982), 'Presidential initiative', *Estates Gazette*, Sept. 25.

Lloyd, M.G. (1984a), 'EZs: the evaluation of an experiment', *The Planner*, June.

Lloyd, M.G. (1984b), 'Policies in search of an opportunity', *Town and Country Planning*, Vol. 53, No.11.

Lloyd, M.G. (1985a), 'Privatisation, liberalisation and simplification of statutory land use planning in Britain', *Planning Outlook*, Vol. 28, No. 1.

238

Lloyd, M.G. (1985b), 'Technology EZs: variation on a theme', *Estates Gazette*, June 22nd.

Lloyd, M.G. (1987), 'Simplified Planning Zones – the privatisation of land use controls in the UK', *Land Use Policy*, Jan.

Lloyd, M.G. and Newlands, D. (1988), 'Business and planning: a privatised planning initiative', *Planning Practice and Research*, No. 4.

Lock, D. (1987), 'The making of Greenland Dock', *The Planner*, March.

Lockard, D. (1971), 'Patent racism', *Trans-action*, Vol. 8, No. 5/6.

London Dockland Development Corporation (LDDC) (1984a), *Corporate Plan, Objectives, Policies and Strategies*, London, LDDC.

London Dockland Development Corporation (1984b), *Proof of Evidence: North Southwark Local Plan Inquiry: Introductory Statement*, London, LDDC.

London Dockland Development Corporation (1984c), *Proof of Evidence: North Southwark Local Plan Inquiry: Policy Objections*, London, LDDC.

London Dockland Development Corporation (1984d), *Proof of Evidence: North Southwark Local Plan Inquiry: Land Use Allocations Objections*, London, LDDC.

London Edinburgh Weekend Return Group (1979), *In and Against the State*, London, LEWRG.

London Planning Advisory Committee (LPAC) (1987), *Review of the Use Classes Order 1987 – Report No. 115/87*, London, LPAC.

Loughlin, M. (1980a), 'Planning control and the property market', *Urban Law and Policy*, Vol. 3, pp. 1–22.

Loughlin, M. (1980b), 'The scope and importance of "material considerations"', *Urban Law and Policy*, Vol. 3, pp. 171–192.

Loughlin, M. (1981), 'Local government in the welfare corporate state', - *Modern Law Review*, Vol. 44.

Loughlin, M. (1982), '"Planning gain": another viewpoint', *Journal of Planning and Environment Law*, June.

Luder, O. (1986), 'The keys to success are cut already', *Building*, May 2nd.

Lyal, S. (1983), 'Building a new London', *New Society*, Jan. 6th.

MacDonald, K. (1983a), 'Shotgun planning at Hays Wharf', *Planning*, July 15th.

MacDonald, K. (1983b), 'Off the fence', *Town and Country Planning*, Vol. 52, No. 10.

Macmillan, H. (1938), *The Middle Way*, London, Macmillan.

McAuslan, P. (1980), *The Ideologies of Planning Law*, Oxford, Pergamon.

McAuslan, P. (1981), 'Local government and resource allocation in England: changing ideology, unchanging law', *Urban Law and Policy*, Vol. 4.

McAuslan, P. (1982), *Law, Market and Plan in the 1980s*, Cambridge, Department of Land Economy.

McIlroy, J. (1983), 'Free for some', *New Statesman*, Apr. 15th.

McKay, D. and Cox, A. (1979), *The Politics of Urban Change*, London, Croom Helm.

McKee, W. (1982), 'Is there a future for planning?', *The Planner*, Feb.

McLennan, G., Held, D. and Hall, S. (eds), *State and Society in Contemporary Britain*, Cambridge, Polity Press.

Bibliography

McLoughlin, B. (1969), *Urban and Regional Planning: A Systems Approach*, Faber, London.

Marks, S. (1983), 'Doubts on reasons for EZ success', *Estates Gazette*, May 7th.

Marshall, T.H. (1964), *Class, Citizenship and Social Development*, New York, Doubleday and Co.

Martyn, N. (1981), 'London's UDC: no respect for democracy', *Town and Country Planning*, Vol. 50, No. 11/12.

Massey, D. (1982), 'Enterprise Zones: a political issue', *International Journal of Urban and Regional Research*, Vol. 6, No. 3.

Ministry of Housing and Local Government (MHLG) (1962), *Town Centres; Approach to Renewal*, London, HMSO.

Milne, R. (1982), 'Why the Dockland Corporation won't release its plan', *Planning*, Dec. 17th.

Milne, R. (1983), 'South Bank City unsure of beating the office blues', *Planning*, Oct 21st.

Mobbs, N. (1983), 'Can development and town planning objectives be reconciled?', *The Planner*, Nov./Dec.

Montgomery, J. and Thornley, A. (eds) (1990), *Radical Planning Initiatives; New Directions for Urban Planning in the 1990s*, Aldershot, Gower.

Newman, I. and Mayo, M. (1981), 'Docklands', *International Journal of Urban and Regional Research*, Dec.

Norcliffe, G. and Hoare A. (1982), 'EZ policy for the inner city: a review and preliminary assessment', *Area*, Vol. 14, No. 4.

Norton, P. and Aughey, A. (1981), *Conservatives and Conservatism*, London, Temple Smith.

Nott, S.M. and Morgan, P.H. (1986), 'Development Plans: what role for the law?', *Journal of Planning and Environnment Law*, Dec.

Nuffield Foundation (1986), *Town and Country Planning. The Report of a Committee of Inquiry*, London, Nuffield Foundation.

O'Connor, J. (1973), *The Fiscal Crisis of the State*, New York, St Martin's Press.

O'Dowd, L. and Rolston, B. (1985), 'Bringing Hong Kong to Belfast? The case of an enterprise zone', *International Journal of Urban and Regional Research*, Vol. 9, No. 2.

Offe, C. (1984), *The Contradictions of the Welfare State*, London, Hutchinson.

Oppenheimers (1987), *The Town and Country Planning (Use Classes) Order 1987*, London, Oppenheimers.

O'Sullivan, N. (1976), *Conservatism*, London, Dent.

Parker, G. and Oatley, N. (1989), 'The case against the proposed UDC for Bristol', *The Planner*, Jan.

Pearce, B. (1981), 'An emerging style of planning', *Planning Outlook*, Vol. 23, No. 1.

Pearce, B. (1984), 'Development control: a neighbour protection service?', *The Planner*, May.

Pearce, B., Curry, N. and Goodchild, R. (1978), *Land, Planning and the Market*, University of Cambridge, Department of Land Economy Occasional Paper No. 9.

Pennance, F. (1974), 'Planning, land supply and demand', in Walters, A. *et al.*, *Government and the Land*, London, Institute of Economic Affairs.

Pickering, M. (1987), 'Development control topics', *The Planner*, Feb.

Picture Post (1941), 'A Plan for Britain', *Picture Post*, Vol. 10, No. 1.

The Planner (1988a), 'Heseltine v. Ridley on the SE', *The Planner Mid-month Supplement*, April.

The Planner (1988b), 'Two views from the Conservative Conference', *The Planner*, Nov.

Planning (1979), 'Rough ride for South Yorks plan from Tories', *Planning*, July 13th.

Planning (1980a), 'More flexibility over change of use', *Planning*, July 11th.

Planning (1980b), 'Members blamed', *Planning*, Aug. 29th.

Planning (1980c), 'Aesthetic control', *Planning*, Sept. 5th.

Planning (1980d), 'Government concessions over GDO relaxations', *Planning*, Nov. 14th.

Planning (1981a), 'Circular begins to bite', *Planning*, Feb. 27th.

Planning (1981b), 'Catch 22 in notorious circular', *Planning*, Mar. 20th.

Planning (1981c), 'Rag-bag of control advice', *Planning*, Apr. 3rd.

Planning (1981d), 'Circular cases support doubts', *Planning*, Aug. 10th.

Planning (1981e), 'Uncertainty over policy on Orders', *Planning*, Nov. 13th

Planning (1982), 'Reacting to housing pressure', *Planning*, Feb. 19th.

Planning (1983a), 'Enterprise houses worry for councils', *Planning*, Mar. 18th.

Planning (1983b), 'More compliance time allowed in 22/80 cases', *Planning*, Apr. 29th.

Planning (1983c), 'Reservations all round in responses to draft planning gain advice', *Planning*, May 13th.

Planning (1983d), 'Development Order beats the eight week limit', *Planning*, July 15th

Planning (1985a), 'Commerce blows a gale on Draft Plan', *Planning*, Feb. 8th.

Planning (1985b), 'DoE wages war on "restrictive" planning', *Planning*, Aug. 30th.

Planning (1985c), 'When 22/80 is cast aside', *Planning*, Aug. 30th.

Planning (1986a), 'Association joins key city council in rejecting Use Classes proposals', *Planning*, Feb. 14th.

Planning (1986b), 'Use Class reform plans get diluted', *Planning*, June 20th.

Planning (1986c), 'Business Class tipped to disappoint users', *Planning*, Nov. 14th.

Planning (1987), 'Local control in the balance with investment race', *Planning*, Mar. 13th.

Plant, R. (1983), 'The resurgence of ideology', in Drucker, H., Dunleavy, P., Gamble, A. and Peele, G. (eds), *Developments in British Politics*, London, Macmillan.

Plant, R. (1984), *Equality, Markets and the State*, London, Fabian Society.

Property Advisory Group (1981), *Planning Gain*, London, HMSO.

Property Advisory Group (1985), *Report on Town and Country Planning (Use Classes) Order 1972*, London, Department of Environment.

Punter, J. (1986), 'The contradictions of Aesthetic Control under the Conservatives', *Planning Practice and Research*, No. 1.

Purdue, M. (1982), 'Commentary on law reports', *Journal of Planning and Environmental Law*, September.

Purton, P. and Douglas, C. (1982), 'EZs in the UK: a successful experiment?', *Journal of Planning and Environment Law*, July.

Pym, F. (1984), *The Politics of Consent*, London, Hamish Hamilton.

Raine, J., Mobbs, T. and Stewart, J. (1980), *The Local Government, Planning and Land Act, 1980, in Perspective*, University of Birmingham, INLOGOV, Dec.

Ravetz, A. (1980), *Remaking Cities*, London, Croom Helm.

Reade, E.J. (1980), *Town Planning and the 'Corporatism Thesis'*, Sociologists in Polytechnics Paper No. 10.

Reade, E.J. (1982), 'Section 52 and corporatism in planning', *Journal of Planning and Environment Law*, Jan.

Reade, E.J. (1987), *British Town and Country Planning*, Milton Keynes, Open University Press.

Rees, G. and Lambert, J. (1985), *Cities in Crisis*, London, Edward Arnold.

Regan, D.E. (1978), 'The pathology of British land use planning', *Local Government Studies*, April.

Riddell, P. (1983), *The Thatcher Government*, Oxford, Martin Robertson.

Ridley, N. (1987), 'Secretary of State's Address to RTPI Summer School 1986', *The Planner*, Feb.

Ross, J. (1983), *Thatcher and Friends: The Anatomy of the Tory Party*, London, Pluto.

Rossi, P. (1971), 'The city as purgatory', *Social Science Quarterly*, Vol. 51, No. 4.

Rowan Robinson, J. and Lloyd, M.G. (1986), 'Lifting the burden of planning: a means or an end?', *Local Government Studies*, May/June.

Royal Institute of Chartered Surveyors (1986), *A Strategy for Planning*, London, RICS.

Royal Town Planning Institute (1986), *Town and Country Planning (Use Classes) Order – Memorandum of Observation to the DoE*, London, RTPI.

Russel, T. (1978), *The Tory Party*, Harmondsworth, Penguin.

Rutherford, M. (1983), 'Review of *Politics of Thatcherism* edited by Hall and Jacques', *Marxism Today*, July.

Rydin, Y. (1986), *Housing Land Policy*, Aldershot, Gower.

Saunders, P. (1979), *Urban Politics; A Sociological Interpretation*, London, Hutchinson.

Saunders, P. (1981), *Social Theory and the Urban Question*, London, Hutchinson.

Schiffer, J. (1983), 'Urban enterprise zones: a comment on the Hong Kong model', *International Journal of Urban and Regional Research*, Vol. 7, No. 3.

Schumpeter, J.A. (1974), *Capitalism, Socialism, and Democracy*, London, Allen & Unwin (first published in Great Britain in 1943).

Scruton, R. (1980), *The Meaning of Conservatism*, Harmondsworth, Penguin.

Select Committee of the House of Lords (1981), *Report on the London Dockland Development Corporation (Area and Constitution) Order 1980*, London, HMSO.

Seyd, P. (1980), 'Factionalism in the 1970s', in Layton-Henry, Z. (ed.), *Conservative Party Politics*, London, Macmillan.

Sharp, T. (1940), *Town Planning*, Harmondsworth, Penguin.

Sherman, A. (1970), 'The end of local government', in Boyson, R. (ed.), *Right Turn*, London, Churchill Press.

Shostak, L. and Lock, D. (1984), 'The need for new settlements in the SE', *The Planner*, Nov.

Shostak, L. and Lock, D. (1985), 'New country towns in the SE', *The Planner*, May.

Shutt, J. (1984), 'Tory enterprise zones and the labour movement', *Capital and Class*, No. 23, Summer.

Siegan, B. (1972), *Land-Use without Zoning*, Lexington, Mass., Lexington Books.

Siegan, B. (1976), *Other People's Property*, Lexington, Mass., Lexington Books.

Simpson, I. (1987), 'Planning gain; an aid to positive planning?', in Harrison M.L. and Mordey, R. (eds), *Planning Control: Philosophies, Prospects and Practice*, London, Croom Helm.

Skidelsky, R. (ed.) (1988), *Thatcherism*, London, Chatto & Windus.

Slough, D. (1974), 'Labour scarcity and costs', in Walters, A. *et al.*, *Government and the Land*, London, Institute of Economic Affairs.

Sorensen, A.D. (1982), 'Planning comes of age: a liberal perspective', *The Planner*, Nov./Dec.

Sorensen, A.D. (1983), 'Towards a market theory of planning', *The Planner*, May/June.

Sorensen, A.D. and Day, R.A. (1981), 'Libertarian planning', *Town Planning Review*, Vol. 52.

Southall, A. (1983), 'Structure plans: legal aspects and problems', in *Structure Plans and Local Plans – Planning in Crisis*, Journal of Planning and Environment Law Occasional Papers, London, Sweet & Maxwell.

Southwark London Borough (1984a), *Planning Newssheet*, Sept. 25th, London, Southwark LB.

Southwark London Borough (1984b), *North Southwark Plan Local Inquiry: Introductory Statement: Proof of Evidence: London Dockland Development Corporation*, London, Southwark LB.

Stansfield, K. (1987), 'UDCs – gearing up for action', *Local Government News*, Oct.

Steen, A. (1981), *New Life for Old Cities*, London, Aims of Industry.

Stungo, A. (1985), 'Simplified planning zones explained', *Estates Gazette*, Dec. 14th.

Suddards, R. (1986), 'Planning law – a wind of change', *The Planner*, Feb.

Swann, P. (1980), 'GDO: a useful relaxation?', *Planning*, Mar. 7th.

Swann, P. (1981), 'Delayed initial strategy for Merseyside disappoints', *Planning*, Sept. 25th.

Swann, P. (1982), 'Getting "community" into "reasonable" planning gain', *Planning*, Feb. 26th.

Swann, P. and Milne, R. (1982), 'Plans docked', *Planning*, June 18th.

Taylor, S. (1981), 'The politics of EZs', *Public Administration*, Vol. 59, Winter.

Bibliography

Taylor-Gooby, P. and Dale, J. (1981), *Social Theory and Social Welfare*, London, Edward Arnold.
Thatcher, M. (1977), *Let Our Children Grow Tall*, London, Centre for Policy Studies.
Thomas, H. (1984), 'The fruits of conservatism', *New Society*, Mar. 22nd.
Thomas, H. (1989), 'Evaluating CBDC's effectiveness', *The Planner*, Jan.
Thompson, R. (1983), 'Powerless to react to plans for wharf world decline', *Planning*, June 17th.
Thompson, R. (1987), 'Is fastest best? – The case of development control', *The Planner*, Sept.
Thompson, R. (1990), 'An achievable alternative for planning', in Montgomery, J. and Thornley, A. (eds), *Radical Planning Initiatives*, Aldershot, Gower.
Thornley, A. (1981), *Thatcherism and Town Planning*, Polytechnic of Central London Planning Studies No. 12.
Thornley, A. (1986), 'Thatcherism and simplified regimes', *Planning Practice and Research*, No. 1.
Thornley, A. (1988), 'Planning in a cool climate – the effects of Thatcherism', *The Planner*, July.
Titmuss, R. (1950), *Problems of Social Policy*, London, HMSO.
Titmus Sainer & Webb and Fuller Peiser (1987), *The 1987 Use Classes Order – does it achieve its aims?*, London, Titmus Sainer & Webb and Fuller Peiser.
Town and Country Planning Association (TCPA) (1980), 'Urban Development Corporations', *Town and Country Planning*, Vol. 49, No. 1.
Town and Country Planning Association (1981), 'Wanted: a plan', *Town and Country Planning*, May.
Town and Country Planning Association (1984), *Simplified Planning Zones*, London, Town and Country Planning Association.
Tullock, G. (1976), *The Vote Motive*, London, Institute of Economic Affairs.
Tullock, G. (1979), *Taming of Government*, London, Institute of Economic Affairs.
Tym, R. (1984), *EZ Monitoring Report – Year Three*, London, Roger Tym & Partners.
Urry, J. (1981), *The Anatomy of Capitalist Societies: The Economy, Civil Society and the State*, London, Macmillan.
Walters, A.A. (1974), 'Land speculator – creator or creature of inflation', in Walters, A.A. et al., *Government and the Land*, London, Institute of Economic Affairs.
Walters, A.A., Pennance, F.G., West, W.A., Denman, D.R., Bracewell-Milnes, B., Denman, S.E., Slough, D.G. and Ingram, S. (1974), *Government and the Land*, London, Institute of Economic Affairs.
Ward, R. (1981), Lecture on London Docklands given at Polytechnic of Central London, Feb. 5th, unpublished.
Ward, R. (1982), 'London's Docklands: the LDDC's aims', *Planner News*, July.
Watson, G. (1983), 'Who are the conservatives?', *Encounter*, Dec.
Webber, M. (1974), 'Permissive planning: planning in an environment of

change', in Blowers, A., Hamnett, C. and Sarre, P. (eds), *The Future of Cities*, London, Hutchinson. This is a reprint of Town Planning Review, Vol. 39, Nov.

West, R.J. (1983), 'Difficulty of control and the Use Classes Order', in *Retail Planning and Development*, London, PTRC.

West, W.A. (1974), 'Town planning controls – success or failure', in Walters, A. *et al.*, *Government and the Land*, London, Institute of Economic Affairs.

Wiener, M.J. (1981), 'Conservatism, economic growth and English culture', *Parliamentary Affairs*, Autumn.

Wilbraham, P. (1982), 'Recent developments in planning law', *The Planner*, Feb.

Williams, R.H. and Butler, P. (1982), 'EZs – dogma abandoned', *Town and Country Planning*, March.

Wingo, L. and Wolch, J.R. (1982), 'Urban land policy under the new conservatism', *Urban Law and Policy*, Vol. 5.

Wray, I. (1986), 'UDCs – hitting the panacea button', *Architects Journal*, Nov. 5th.

Wray, I. (1987), 'The Merseyside Development Corporation: progress versus objectives', *Regional Studies*, Vol. 21, No. 2.

Index

Milton Keynes UK
Ingram Content Group UK Ltd.
UKHW031147141024
449569UK00024B/1007